Losing Touch with Nature

Losing Touch with Nature

Literature and the New Science in Sixteenth-Century England

MARY THOMAS CRANE

Johns Hopkins University Press
Baltimore

Printed in the United States of America on acid-free paper
2 4 6 8 9 7 5 3 1

Johns Hopkins University Press
2715 North Charles Street
Baltimore, Maryland 21218-4363
www.press.jhu.edu

Library of Congress Cataloging-in-Publication Data

Crane, Mary Thomas, 1956– author.
Losing touch with nature : literature and the new science in sixteenth-century England / Mary Thomas Crane.
p. cm.
Includes index.
ISBN 978-1-4214-1531-4 (hardcover : acid-free paper) — ISBN 978-1-4214-1532-1 (electronic) — ISBN 1-4214-1531-3 (hardcover : acid-free paper) — ISBN 1-4214-1532-1 (electronic) 1. English literature—Early modern, 1500–1700—History and criticism—Theory, etc. 2. Literature and society—England—History—16th century. 3. Literature and science—England—History—16th century. 4. England—Intellectual life—16th century. I. Title.
PR418.S64C74 2014
820.9'003—dc23 2014006749

A catalog record for this book is available from the British Library.

Special discounts are available for bulk purchases of this book. For more information, please contact Special Sales at 410-516-6936 or specialsales@press.jhu.edu.

Johns Hopkins University Press uses environmentally friendly book materials, including recycled text paper that is composed of at least 30 percent post-consumer waste, whenever possible.

For Thomas Crane and Parker Crane

Contents

Acknowledgments

A part of chapter 4 appeared in an earlier form as "Spenser's Giant and the New Science," *Go Figure: Energies, Forms, and Institutions in the Early Modern World*, ed. Judith H. Anderson and Joan Pong Linton (New York: Fordham University Press, 2011), 19–37. Part of chapter 5 was published in a previous version as "The Physics of *King Lear*: Cognition in a Void," Graham Bradshaw, Tom Bishop, and Mark Turner, eds., *The Shakespearean International Yearbook* 4 (2004): 3–23. Parts of chapter 6 appeared as "Marlowe and the New Science," *Christopher Marlowe in Context*, ed. Emily Bartels and Emma Smith (Cambridge: Cambridge University Press, 2013), 252–61, and as "Roman World, Egyptian Earth: Cognitive Difference and Empire in Shakespeare's *Antony and Cleopatra*," *Comparative Drama* 43 (2009): 1–18. Reprinted in Shankar Raman and Lowell Gallagher, eds., *Knowing Shakespeare: Senses, Embodiment, and Cognition* (New York: Palgrave, 2010), 102–15.

This book took too long to write. As happens to many of us, I have spent the past ten years serving as a department chair (for two different departments) and doing other administrative work. On the one hand, I think the book may be better than it would have been had I been able to write it more quickly since I had more time to think about a range of possible objections to its argument and to read more deeply in relevant work in the history of science. On the other hand, there are now as a result of this extra time too many notes. I hope they will at least serve as a resource for literary scholars pursuing similar topics.

I was sustained during these years by reading the work of, and especially by talking to (and arguing with), a number of scholars also researching topics related to early modern literature and science. Seminars at the annual conference of the Shakespeare Association of America and panels at the annual conferences of the Modern Language Association and the Renaissance Society of America have been the site of especially fruitful dialogues. I'm sure I will leave someone out, but I have benefited from the work of Piers Brown, Katherine Eggert, Wendy Hyman, Rebecca Lemon, Scott Maisano, Howard Marchitello, Carla Mazzio, Gail Kern Paster, Kristen Poole, Debapriya Sarkar, Douglas Trevor, Rebecca Totaro, Evelyn Tribble, and Travis Williams. I also profited from discussion at "Radical Readings: A Workshop on Early Modern Artifacts," organized at Boston College by Sarah Gwyneth Ross and Virginia Reinburg. Exchanges with Paula Blank, Elizabeth Spiller, and Mary Floyd-Wilson were particularly useful to me. To my regret, Floyd-Wilson's excellent book, *Occult Knowledge, Science, and Gender on the Shakespearean Stage,* appeared in print too late to be fully incorporated into my argument here. My neighbors at the Massachusetts Institute of Technology, Diana Henderson and Shankar Raman, have helped with probability and mathematics. Shankar read the book for the press and provided his usual very helpful, exhaustive comments. I have enjoyed working with Henry Turner as coeditor of our Ashgate series on the Literary and Scientific Cultures of Early Modernity and have learned a lot from him.

My thinking on the cognitive implications of this topic has also been furthered by my participation in the Balzan project "Literature as an Object of Knowledge," organized by Terence Cave at the St. John's College Research Centre, Oxford. I have had especially helpful conversations with Terence, Kathryn Banks, Raphael Lyne, Tim Chesters, Miranda Anderson, Guillemette Bolens, and Karin Kukkonen. In the United States, cognitive critics Michael Booth, Amy Cook, F. Elizabeth Hart, and Lisa Zunshine have influenced my thinking. I have also been urged along the cognitive path by the inimitable Ellen Spolsky and by my colleague Alan Richardson.

At Boston College, I need to thank the Classics Department, Kendra Eshleman, Gail Hoffman, and Brigitte Libby, for basically running themselves while I pretended to be their chair, and for answering my questions about Latin. I'm also grateful to my early modern colleagues in the English Department at Boston College, Dayton Haskin, Andrew Sofer, Caroline Bicks, and especially Amy Boesky, who has been a good friend for many years. Several graduate students provided research assistance for this project over the years, and I'm grateful to Jaime Goodrich, Katherine Kellett, Alice E. Waters, Emma Atwood, and Deanna Malvesti for their help with this and other things. Deans David Quigley and Andy Boynton, deserve thanks for helping me explore the costs and benefits of irrational persistence. My former colleagues Ti Bodenheimer and Andy Von Hendy have been a source of encouragement, as always. Finally, I am grateful to my colleague Laura Tanner for her work on our joint project to stamp out germs and encourage professionalism.

Now that I'm old, I'm grateful that I still have some very old friends. Teresa D'Orazio and Elena Calhoun date back to Spotsylvania High School. I'm lucky that Nita Sembrowich is still living in Cambridge and that we're proceeding with our goal of becoming little old Cambridge ladies. From graduate school, I'm still happy to have the friendship of Laura Knoppers and to be able to rely on the advice of Barbara Lewalski. I don't know what I would do without regular telephone chats with Emily Bartels, especially for commiseration about administrative challenges.

My sons Tom and Parker are grown up now and were hardly inconvenienced at all by the writing of this book. It is dedicated to them.

Losing Touch with Nature

CHAPTER 1

Introduction

In recent years, to fill in the areas of history neglected by New Historicism, a number of scholars have turned their attention to the relationship between the scientific revolution and the literary and cultural artifacts of the late sixteenth and early seventeenth centuries. Building on new developments in the history of science—especially the work of scholars like Steven Shapin, Simon Schaffer, Peter Dear, Katharine Park, Lorraine Daston, Barbara Shapiro, and Richard Westfall, who delineate the social and cultural embeddedness of scientific discovery in the period—this valuable, ground-breaking scholarship has greatly enlarged our view of the formative influence of humanist habits of thought on the development of the new science and of the contributions of literary and theatrical forms to new epistemologies and new structures of knowledge.[1] Literary scholars have also been influenced by historians of science who have newly emphasized the contributions of artisans and practitioners, rather than university-trained philosophers, to the development of the new science, and they have convincingly shown how literary "makers" participated in the construction of new forms of knowledge.[2] This literary scholarship has tended to look forward to the seventeenth century, to empirical methods and new ideas, and it has seen signs of these nascent in sixteenth and early seventeenth century literature.[3]

Historians of science are aware of the dangers of taking an overly teleological view of their subject—of writing "Whig history"—and acknowledge that they are often tempted to look for signs of progress, of movement toward new ideas, of "getting it right." In doing so they blind themselves to the wrong ideas and lingering beliefs that, paradoxically, ended up providing a way forward toward scientific revolution.[4] As much as they acknowledge this methodological problem and consciously avoid it, the fact remains that as historians of *science*, their project is ultimately to chart the development of ideas toward newer forms. Even William Newman, who has argued passionately that alchemy should be taken seriously on its own terms, is nevertheless largely interested in its ultimate contributions to the development of modern chemistry.[5]

Literary scholars who have attended to the relationship between science and literature in this period have also been aware of the dangers of "scientism" but have nevertheless tended ever so slightly toward wanting literature to be on the side of the new, rather than the outmoded, and to see humanistic disciplines as contributing to the scientific revolution. What would be the point of showing that early modern literature was backward looking? It has been much more satisfying to show our colleagues in the sciences that they could not have gotten where they are without the foundational contributions of literary texts and humanist thinkers.[6]

My interest in this book, however, is in the period of ferment, confusion, and angst between around 1530 and 1610 in England, when the settled Aristotelian, Galenic, and Ptolemaic accounts of how the universe worked began to fall apart and the new ideas that would replace them were still inchoate and in flux. Educated people in England knew enough to understand bits and pieces of the new ideas emanating from the continent and from some thinkers in England but did not have a sense of what they added up to. Even experts, like the members of the Dee circle who had advanced mathematical training, were struggling to figure out how developments like Copernican astronomy, advances in mathematics and optics, particulate theories of matter, and germ theories of disease were to be reassembled into a coherent and convincing whole. This group bridges the theory/practice divide: they were involved in practical projects but also addressed issues of epistemology and methodology. The experts in the Dee circle and educated Londoners familiar with their work were aware that an unprecedented epistemological rupture was under way, a sundering of an embodied, intuitive grasp of the world from the specialist's increasingly abstract accounts of how things really worked. These educated people included writers like Christopher Marlowe, William Shakespeare, and Edmund Spenser, and some of the most important literary works produced in England during this period were crucially shaped by the mingled elation and horror occasioned by the changes their authors were living through—elation at the new power afforded by abstraction and a newly malleable universe; horror at the idea that lived experience of the world did not reliably provide access to truth about it.

This book is a cognitive history in the sense employed by Reviel Netz: a "study of culturally specific practices, in which universal human cognitive abilities are assembled together and implemented with the aid of specific tools and technologies."[7] It is also a cognitive history in a slightly different sense because it discusses how the loss of the Aristotelian view of the world *felt* to English writers in the second half of the sixteenth century: "felt" in the sense of both perception and affect, grounded in the cognitive insight that embodied perception, affect, and rational thought are inextricable.[8] I might describe this as a reception history (or prehistory) of the scientific revolution, and it is based in a sustained (although necessarily partial) attempt on my part to understand what these writers knew from their perspective and to reimagine their feelings. I have tried to follow William Newman's directive to base my account on "a close reading of the original primary texts with all their recalcitrant vexations."[9] Literary texts have a role here not so much because they advanced scientific knowledge but because they registered, in particularly complex ways, the welter of ideas and feelings that circulated in response to changing views of the world.

Research in cognitive psychology has suggested that human experience of embodiment provides us with powerful "intuitive" concepts about the way the universe works, concepts that are so tenacious that they survive college-level instruction and persist in the work of highly advanced mathematical thinkers.[10] *Intuition* here is defined as concepts that seem "self-evident," that are developed without a rational process of thought, are held with great confidence, and persevere in the face of evidence to the contrary.[11] Studies have shown that even students who have taken a year of college physics still fail to grasp basic Newtonian laws of force and motion and retain their incorrect intuitive ideas about the behavior of objects in the world.[12] Susan Carey has identified the dilemma of science educators: "our scientific heritage has provided us with deep and counterintuitive understanding of the physical, biological, and social worlds," and it is difficult to get students to replace this understanding of the world with counterintuitive models.[13]

I was struck, then, when I first came across the assertion that some Aristotelian theories about the world closely resemble the intuitive concepts that make science education difficult today. Surely the divergence of official "science" from intuitive human understanding of the world constituted a significant rupture in the sixteenth and seventeenth centuries.[14] I encountered the relationship between Aristotle and intuitive science for the first time in the work of Joachim Büttner, Peter Damerow, Jürgen Renn, Matthias Schemmel, and Matteo Valleriani of the Max Planck Institute for the History of Science in Berlin, who argue that "the most basic knowledge presupposed by physics is based on experiences acquired almost universally in any culture by human activities such as moving one's body around or handling objects under normal conditions characteristic of our natural environment on earth."[15] They suggest that these "fundamental notions" do not correspond to the categories of Newtonian mechanics but instead "resemble the basic assumptions of Aristotelian physics."[16]

Newton's concept of inertia, for instance, coincides with intuitive physics in the case of objects at rest, which need an application of force to be moved. But Newton's law diverges from intuition in the case of objects in motion, since it holds that an object in motion will continue moving in a straight line indefinitely unless a force is applied to stop it. Friction, the invisible force that stops objects from moving, cannot be directly observed, so our intuitive sense is that objects in motion need an application of force in order to continue moving. Aristotelian mechanics posited such a force, called *impetus*, which was thought to move objects along until it was exhausted. Nancy Nersessian notes that "the concepts students intuitively employ to understand how objects move resemble more the Aristotelian/medieval concepts than the Newtonian understanding needed to acquire the science, which is most likely due to their experiences in a world of friction. And studies show that the intuitive conceptual structure is largely untouched by traditional science instruction."[17]

Other scholars have identified additional aspects of Aristotelian physics with intuitive beliefs. Steven Shapin notes that the seventeenth-century atomists' "distinction between primary and secondary qualities, just like the Copernican view of the world, drove a wedge between the domain of philosophical legitimacy and that of common sense," and upon further reflection it becomes clear that many aspects of Aristotelian, Ptolemaic, and Galenic philosophies are elaborations of intuition.[18] We now take for granted the ideas that the earth revolves around the sun, that disease is spread by invisible organisms, and that seemingly solid matter is composed of tiny particles, but all of these concepts are counterintuitive, at odds with our embodied experience of the world, and equally problematic within the Aristotelian system.

The congruence between these ancient theories and intuitive science is not an accident but rather a function of the epistemological basis of Aristotelian natural philosophy in ordinary perceptual experience of the world. As Peter Dear has established, Aristotelian natural philosophy was grounded in "experience," which is defined as "a universal statement about how things are, or how they behave."[19] Experience was based on observation of nature but was sharply distinguished from "experiments," which involve singular (and artificially induced) events. The four elements of Aristotelian terrestrial physics were defined by their sensible qualities so that knowledge about them could only come from observation of their behavior as manifested to human sensory experience.[20] For Aristotelian natural philosophers, truth about nature could only come from reference to "a behavior or appearance in nature that happened routinely and with which everyone was familiar from daily experience," that is, from intuitive knowledge of the world.[21]

If intuitive science is so deep-seated and persistent even today, what must it have been like to live through the severing of scientific theory from the bedrock of ordinary embodied experience? Several theorists of scientific knowledge have registered the violence of this separation. Michel Serres writes powerfully about the contrast between abstraction and bodily knowledge: "To abstract means to tear the body to pieces rather than merely to leave it behind: analysis."[22] Gaston Bachelard refers to the divergence of scientific thought from ordinary phenomenal experience as "brutal."[23] This book attempts to recapture the early modern loss of an intuitive connection with nature, tracing it through textbooks and treatises published in sixteenth-century England, as well as in some literary works from the period. As we know, there were many reasons for early modern English subjects to have felt an uneasy mix of exhilaration and anxiety: new opportunities for social mobility (up and down), increasing contact with different cultures, a radical unsettling of religious institutions and beliefs, and unprecedented access to education, public dramatic performances, and printed texts. There is clearly no single master narrative that explains the ferment of the Elizabethan and early Jacobean periods, and I don't claim that the epistemological rupture that I trace here provides such a narrative.

Nevertheless, I believe that the slow erosion of the Aristotelian worldview had a greater direct effect on culture and literature in sixteenth-century England than we have acknowledged. Tracing that effect yields an account of the prehistory of the scientific revolution in England, and its influence on literary works, that differs in important ways from some of the narratives that literary scholars have tended to rely on. It also yields an account that differs in some modest ways from the understanding of historians of science, largely because of my emphasis on popularizing treatises and textbooks rather than cutting-edge science, although most of what I argue here is based on work in the history of science that in some cases has not yet been fully assimilated by nonspecialists.[24] In order to argue that the new science destroyed a direct, intuitive connection with nature, it is first necessary to show that sixteenth-century natural philosophy maintained such a connection. Historians of science like Newman and Shapin, along with Keith Hutchison, Margaret Osler, Christoph Meinel, and others, have revealed how advocates of the new science in the seventeenth century misrepresented the Aristotelian position on observation of nature as well as their own ability to provide access to empirical evidence for mechanistic theories.[25] Other scholars have emphasized previously unrecognized continuities between Aristotelian theories and the new science.[26] Aristotelian "experience" can, in a way, be seen as involving a more direct connection with observation of the world than did empiricism, which had to accept indirect and probabilistic evidence for phenomena that were not directly observable.

As I read work by these contemporary historians of science and also textbooks and treatises from the early modern period, I realized that I had come to feel that the Aristotelian account of the world made perfect sense. The behavior of the four elements, the meteorological theories that explained a range of natural phenomena as produced by vapors drawn up out of the elements by the action of the sun, even the argument that if the sun and moon clearly influence the operations of the elements on earth, then surely the stars exercised a similar, if less detectable, influence—these ideas began to seem as plausible as, perhaps even more plausible than, modern scientific explanations. I found myself trying to explain to people that Dee's angel conversations could be seen as a largely rational attempt to gain direct knowledge about the universe and was greatly relieved to find that Deborah Harkness had reached more or less the same conclusion.[27] If we can learn to suspend our knowledge of modern science, it is possible to recapture something of the intuitive force of premodern natural philosophy.

I was interested to learn, fairly late in my research, that Thomas Kuhn experienced a similar shift into an Aristotelian worldview, and that he based his theory of paradigm change in *The Structure of Scientific Revolutions* on what struck him as a powerful cognitive experience. Kuhn describes a sudden moment of revelation, rather like a conversion experience, when he understood Aristotelian physics on its own terms for the first time, a kind of scientific devolution that he extrapolated forward to create his theory of scientific revolution. As Kuhn describes the moment of discovery, his insight is inextricable from strong feeling and from embodied experience. Kuhn had been puzzling over the fact that Aristotle's theories of motion and mechanics were, from a Newtonian perspective, so completely wrong. He wondered how these theories had "been taken seriously for so many centuries after his death" given how wrong they were. It occurred to Kuhn that "the fault" might be "mine rather than Aristotle's. . . . Perhaps his words had not always meant to him and his contemporaries quite what they meant to me and mine."[28] His description of the moment of sudden illumination follows:

> Feeling that way, I continued to puzzle over the text, and my suspicions ultimately proved well-founded. I was sitting at my desk with the text of Aristotle's *Physics* open in front of me and with a four-colored pencil in my hand. Looking up, I gazed abstractedly out the window of my room—the visual image is one I

> still retain. Suddenly the fragments in my head sorted themselves in a new way, and fell into place together. My jaw dropped, for all at once Aristotle seemed a very good physicist indeed, but of a sort I'd never dreamed possible. Now I could understand why he had said what he'd said, and what his authority had been. Statements that had previously seemed egregious mistakes, now seemed at worst near misses within a powerful and generally successful tradition. That sort of experience—the pieces suddenly sorting themselves out and coming together in a new way—is the first general characteristic of revolutionary change that I shall be singling out after further consideration of examples.[29]

Kuhn describes a moment of *einfuhlung*, of literally feeling his way into Aristotle's world: "feeling that way, I continued to puzzle over the text." He remembers the moment of insight as strikingly embodied: "sitting at my desk," "I gazed abstractedly out the window of my room," "the visual image is one I still retain," "my jaw dropped." He imagines his cognitive process in concrete terms: "the fragments in my head sorted themselves in a new way, and fell into place together."

Kuhn spent the rest of his life trying to define and defend in abstract terms his intuition that scientific "paradigms" always change in this kind of sudden, wholesale, shift. Kuhn struggled to define his central terms, "paradigm" and "incommensurability," in response to criticism that they were ill defined in *The Structure of Scientific Revolutions*.[30] Cognitive psychologist Nancy Nersessian describes Kuhn's lifelong difficulty with his terminology as stemming from the intuitive nature of his sudden insight. She notes that Kuhn himself "claimed to have relied too much on his own phenomenal experience as a historian attempting to understand Aristotle's worldview" and that he later turned to cognitive science out of "his need to find a way to articulate this intuition" that scientific paradigms shift suddenly.[31] Kuhn acknowledged that his concept of incommensurability was problematic, "in part due to the role played by intuition and metaphor" in his initial formulation.[32]

It seems likely to me that Kuhn's sudden and powerful intuition about scientific revolutions was primarily an experience of intuition itself: he suddenly understood what it felt like to regain an intuitive physics. His theory of scientific revolution was thus most accurate as a description of what happened in the sixteenth and seventeenth centuries but was partially applicable to subsequent scientific discoveries because science has become progressively more counterintuitive over time. Many scientific advances, especially in physics, have replicated to a lesser extent the initial loss of intuitive certainty, as Bachelard has argued: "scientific progress always reveals a break [rupture], constant breaks, between ordinary [*com-*

mune] knowledge and scientific knowledge."[33] Despite Kuhn's desire to be seen as a philosopher rather than a historian, his greatest contribution may well have been his prescient grasp of the felt cognitive experience of a crucial historical moment.[34]

If the loss of an intuitive relationship with the phenomenal world was such a catastrophic occurrence in late sixteenth-century England, why have scholars not recognized it? Literary and cultural scholars of the period have certainly traced religious, philosophical, political, economic, and social causes for widespread anxiety about change and loss of certainty. The inconsistencies in Book V of *The Faerie Queene* or the nihilism of *King Lear*, which I connect to the loss of intuitive science, have tended to be read in the context of Irish colonialism (in the case of Spenser) and a range of political and economic changes (Shakespeare). These explanations are surely correct. But both texts also reveal concerns about, in Spenser's case, the erosion of Ptolemaic astronomy and, in Shakespeare's case, new ideas about the nature of nothingness in mathematics and in theories of matter.

Spenser's references to Copernican theory and to the supernova of 1572 have remained unremarked partly because they take unexpected forms. His allegorical description of the phenomenon of equinoctial precession has not been linked to the new science because precession (the apparent movement of the signs of the zodiac in a thirty-thousand-year cycle) had been recognized from antiquity. However, Spenser's references to it in Book V and in the *Cantos of Mutabilitie* reveal that contemporary discussion of the location of the supernova above the moon and of the apparent retrograde motion of the planets caused him to take the unprecedented view that precession was an alarming sign that the upper reaches of the cosmos were subject to change and decline. For reasons of literary decorum, Spenser alludes to new ideas indirectly, making them congruent with his poetic narrative by translating them into a mythological allegory.

We know, of course, that John Donne was aware that some people were upset by the loss of a settled, intuitive understanding of the world because he says so directly:

And new Philosophy calls all in doubt,
The element of fire is quite put out;
The sun is lost, and th'earth, and no mans wit
Can well direct him where to looke for it.
And freely confesse that this world's spent,
When in the Planets, and the Firmament
They seeke so many new; they see that this

Is crumbled out againe to his Atomies.
'Tis all in peeces, all cohaerence gone;
All just supply, and all Relation. (ll. 205–14)

Donne wrote this in 1611, and it explicitly expresses dismay that the new philosophy has destroyed the long-standing consensus about the nature of the universe. The Aristotelian understanding of the universe is "in doubt"; the sun and earth, no longer thought to behave as they appear to behave, are "lost"; the traditional elements (which were known by their perceptible qualities) have been replaced by "Atomies." The ideas and feelings that Donne expresses here were not new in 1611. Spenser alluded to new astronomical ideas in *The Faerie Queene* to make a similar point: that the universe has experienced decline since it was created. What is new in Donne is not awareness of, or dismay at, the new philosophy, but his willingness in this strange poem to violate literary decorum on several levels and to refer directly to scientific ideas, using scientific terms, in a poem.

I argue, therefore, that major sixteenth-century English literary works by Spenser, Marlowe, and Shakespeare were reacting directly to the loss of an intuitive connection with nature. Looking first at treatises and textbooks published in England throughout the sixteenth century, I show what some people knew and felt about new ideas in natural philosophy and trace their awareness of and concern about epistemological issues. I follow patterns of language and tropes found in these treatises and show how they also can be found in the writings of Gabriel Harvey, Spenser, Marlowe, Shakespeare, and others. It is possible to identify changing responses to the erosion of intuitive knowledge, shifting almost decade by decade from the 1520s to the end of the century. We can recognize traces of these new ideas in literary texts if we realize that their authors' knowledge of them was piecemeal, took sometimes surprising forms, often involved misunderstanding of crucial concepts, and was often expressed indirectly or metaphorically.

My narrative about the development of knowledge in sixteenth-century England differs in several ways from the familiar account of a text-based authoritative Aristotelianism replaced in the seventeenth century by direct empirical observation of nature, and I am conscious of needing to convince readers that, in England at least, the new science developed somewhat differently than has been generally understood. Literary scholars have been aware that seventeenth-century empiricism emerged in part from the work of artisans and practitioners, but they have been less cognizant of a central strain of Aristotelian naturalism in England that directly addressed epistemological and theoretical issues. Although many writers in this period evince an uncritical belief in magic and in a universe replete with mystical correspondences, this naturalistic Aristotelian tradition had a fairly sophisticated understanding of magic and made only cautious use of elaborated correspondences.[35]

Fundamental to my argument is a more restrictive definition of what I mean by *Aristotle*. Many scholars have, understandably, tended to base their understanding of what Aristotle meant to early modern thinkers on the university curriculum as described by scholars like Charles Schmitt.[36] They emphasize his theories of hylomorphism and "substantial forms" found in the *Physics* (and scholastic elaborations of that work), ideas about the composition of matter that were, later in the seventeenth century, ridiculed by Francis Bacon and by members of the Royal society like Robert Boyle. Stephen Toulmin and June Goodfield have noted that "from our distance it is natural enough to accept Bacon's assessment of the situation, contrasting a ramshackle medieval tradition built from Aristotelian bricks with a modern scientific construction rooted firmly in the *terra firma* of experience. Yet the truth was more complicated."[37] William Newman has criticized historians of science who "have accepted the frequently self-serving rhetoric of the heroes of modern science, themselves intent on portraying their discoveries as radically new."[38]

The seventeenth-century critics of Aristotelianism were to some extent caricaturing their predecessor's ideas in an attempt to dismiss them. Hutchison has shown, for instance, that Walter Charleton's critique of what he termed "that ill-contrived [Aristotelian] sanctuary of ignorance, called occult qualities" has been misread by modern scholars. It represents not a rejection of occult qualities per se, since he ultimately implies that all qualities are occult: "Charleton is not attacking occult qualities but the Aristotelians."[39] Steven Shapin similarly describes how seventeenth-century "ridicule" of the doctrine of substantial forms involved "reiterated insistence that their explanations were uniquely intelligible" and relied on an assertion that the concept of substantial forms "was central to Aristotelian natural philosophy" because of its "domination by priests, who used notions of substantial forms, separated essences, and incorporeal substances to grab a share of state power, to frighten the masses and keep them in awe."[40]

Although the university-educated were certainly versed in Aristotle's *Physics* and did call on concepts like form, substance, accident, and the four causes when they were useful, a more common understanding of the structure of the universe in England in this period was based on what historians of science term Aristotelian *naturalism*, the basic account of the four elements, their qualities, and the differing physics of the sub- and supralunar realms found in works like the *Meteorology*, *On Generation and Corruption*, and *On the Heavens*.[41] As Newman has argued, an alternate theory of matter, based in Aristotle's "more empirical and less given to abstract theorizing" works such as the *Meteorology*, was developed in the Middle Ages alongside hylomorphism and survived, partly in alchemical texts, to influence natural philosophers in the sixteenth and seventeenth centuries.[42] Ptolemaic astronomy and Galenic medicine are compatible with Aristotelian naturalism and are in many ways elaborations of it.

Craig Martin has argued that Aristotle's *Meteorology* was more central to early modern natural philosophy than has been recognized. He shows how meteorology did not share the epistemological assumptions of works like the *Analytics* but was instead based on "natural signs, experience, and observation," which led to "hypothetical determination of the causes of effects." Renaissance meteorological works were informed by "actual weather of the sixteenth century" and, "unlike many parts of Aristotelian natural philosophy," proved to be compatible with new ideas (like corpuscular theories of matter) and methodologies because meteorology was "conjectural, based in experience, and primarily concerned with matter and its motion." Knowledge of meteorology was widespread, since it was taught in the universities and was more accessible than works like the *Physics*. It was accessible partly because it was intuitive and fit with "basic experience of clouds, fog, and mists." It also spanned theory and practice since knowledge about the weather was of practical concern to sailors, farmers, and architects.[43]

Toward the end of the sixteenth century, as challenges to Aristotelian explanations of natural phenomena accumulated, a number of writers supplemented naturalism with ideas derived from various traditions that offered access to secret or hidden phenomena. These "secret" traditions spanned technologies traditionally associated with the occult (astrology, alchemy, Neoplatonism, Hermeticism) and with modern science (mathematics, optics) and provided ways for early modern thinkers to work through the growing gap between what was manifest and what was hidden in nature.[44] English writers as different as Robert Recorde, William Fulke, Harvey, Marlowe, Spenser, and Shakespeare—whatever other philosophical systems they may have referenced—all shared a basic intuitive theory of nature grounded in naturalistic works like the *Meteorology*, and they struggled with the erosion of this theory toward the end of the sixteenth century.

There are a set of modern scholarly misconceptions—what we might call, after Francis Bacon, contemporary idols—that have tended to obscure what (and how) sixteenth-century English subjects thought about their world, especially the role of intuitive natural philosophy. The first of these is "the Myth of the Powerful Book," a tendency to overestimate the authority of ancient texts in the period, to look for the origins of new ideas in old books, and to discount the role of observation of natural phenomena in intellectual change. In part, our tendency to look to books for evidence of what early modern people thought or knew stems from continued belief in the now-questioned idea that their knowledge was entirely based on textual authorities. In addition, as literary scholars we look primarily to texts for evidence, so we have a natural inclination to want to trace the genealogy of ideas as they are passed from one text to another. Indeed, almost all of the evidence about the early modern period that we possess is textual. We love books and understandably want to believe that they can transform the world. Steven Greenblatt's *The Swerve* provides a recent example, arguing that many of the concepts and beliefs on which modernity is based were found in Lucretius's Epicurean poem *De Rerum Natura*, which was rediscovered and preserved through the tenacious and dedicated work of a single scholar, Poggio Bracciolini.[45] Certainly, the rediscovery of Lucretius's poem did inspire a number of early modern writers. Greenblatt rightly stops short of arguing that modern atomism developed from the reading of Lucretius alone, but other literary scholars have been willing to make stronger claims about the influence of a Lucretian (or Ovidian) source for atomism in the early modern period.[46]

As historians of science have shown, while early modern atomism was influenced by several ancient theories (Lucretian Epicureanism as well as Aristotelian ideas about *minima naturalia*), it emerged equally from the observed behaviors of matter that could not be easily explained by the Aristotelian elements or Paracelsan principles.[47] Ancient theories of atomism in some cases inspired seventeenth-century thinkers to adopt corpuscular theories of matter but were also sometimes used to provide a corollary or justification for these new, counterintuitive theories, much as Copernicus cited the heliocentric theory of Aristarchus of Samos to back up his novel system.[48] While the example of Aristarchus may have made his new ideas more thinkable, it was not their primary genesis; he developed his new theory of the universe through mathematical calculations.

When seventeenth-century natural philosophers accused their predecessors of basing their conceptions of nature on the authority of Aristotle rather than on observation of nature itself, they were exaggerating the situation for polemical reasons.[49] In the first place, the experimental empiricism they were advocating usually involved not direct observation of nature itself but the construction of analogies between artificial experimental conditions (the vacuum in an air pump, for example) and the ordinary course of nature. Aristotle's works (as well as those of Galen and other ancient writers) certainly possessed immense cultural authority, but their accounts of the natural world did not hold sway for so long simply because of that authority. As I have been suggesting, many of Aristotle's naturalistic arguments retained their force because they corresponded with intuitive experience of the world. Peter Dear has usefully clarified the role of ancient texts in medieval and early modern scientific argument: "Because the text [of Aristotle] determined the character and function of statements of universal experience, the authority of such statements derived from the place they held in the text and the place they held in natural philosophical inquiry, rather than on blind faith in the assertions of Aristotle."[50] These texts recorded what was accepted as ordinary experience of nature, and their authority is a function of their acceptance as a codification of universal embodied experience over time. The mathematician Robert Recorde warned against blind acceptance of Aristotelian authority as early as 1556, and other English writers signaled a similar willingness to question the arguments of authoritative texts and to favor observation and reason over textual authorities.

Our modern overemphasis on the textual transmission of ideas has sometimes led scholars to view English sixteenth-century epistemological questioning as a rediscovery of ancient skepticism.[51] Various scholars have attributed Shakespeare's interest in the discrepancy between appearance and reality to his reading of Montaigne, and they thus connect Shakespeare through Montaigne to the ancient skeptical traditions. However, as William Hamlin acknowledges, "Shakespeare and his fellow dramatists certainly had the potential to be skeptical without reading Montaigne, and Montaigne had the same potential without reading Sextus or Cicero."[52] Ellen Spolsky has argued from a cognitive perspective that skepticism is a permanent condition of the embodied mind caused by awareness of "the competence of the mind/brain and the local manifestation of that competence."[53]

Certainly, skeptical texts were rediscovered in this period and had an influence on theories of knowledge. Richard Popkin's *History of Scepticism* reveals that Italian and French writers were interested in a revival of ancient skepticism before those ideas spread in the seventeenth century to England, where they were controversial.[54] But if we move too quickly to label all emerging doubts about access to truth in this period as offshoots of academic or Pyrrhonic skepticism, we risk underestimating their influence. Many sixteenth-century English writers were surprised and even shocked by the opening of a gap between ordinary experience and truth. They did not view this gap as something already worked out in antiquity but as a new phenomenon, possibly indicating a state of decline in nature or the impending end of the world. Rather than viewing this rupture as a settled condition, these writers tended to view the loss of faith in intuitive knowledge as newly emerging in the period. They were usually searching desperately for a way to repair the rupture, hoping, for instance, that research in optics could explain why the universe appears to us other than as it actually is or that various "secrets" traditions could provide technologies that would once again enable accurate perceptions of the world.

In order to understand the relationship between intuitive naturalism and other early modern ideas about the world, we need to attend more closely to the audiences for whom texts were written and the contexts in which they were written. Another idol might be called "the myth of unified knowledge," the idea that early modern religious, literary, and natural philosophical accounts of nature were all congruent and added up to a unified world view and that anything that any early modern writer said about the nature of the world can be taken as representing "what people thought." Similarly, as Nicholas Clulee has pointed out, scholars tend to want to associate a figure like John Dee with a "coherent philosophy," and so they identify "a unifying philosophy" like Neoplatonism or Hermeticism "thought to inform Dee's writings," derive it "in an abstract and *a priori* fashion," and apply it to his work.[55]

Of course, as many scholars have argued, early modern philosophical, religious, and even literary discourse cannot be definitively separated from something anachronistically called "science," and the disciplines into which knowledge was divided were not as discrete as they are today.[56] Still, certain kinds of discrimination can and should be made. William Newman and Anthony Grafton have debunked the view, propagated by scholars like Keith Thomas, Brian Vickers, and Frances Yates, that the so-called "occult sciences," including astrology, alchemy, and cabala, represented a single "unified system of belief."[57] They argue that if "we withdraw our minds from the modern cultural stereotype of 'the occult sciences,' it is not immediately obvious that the two fields [astrology and alchemy] shared a closely related theoretical framework."[58] Each early modern field of theory and practice needs to be understood on its own terms, and it is important to view traditional Aristotelian as well as more esoteric fields in this critical way. It is also important to remember that a critique of the Aristotelian university curriculum or of the doctrine of "substantial forms" does not entail a rejection of every aspect of Aristotelian thought. It is important to consider whether the audience for a particular work was imagined as primarily interested in theological questions, rhetorical questions, or natural questions, even though there is overlap among these broad areas. A highly rhetorical and marginally coherent work like Thomas Nashe's *The Terrors of the Night* (1594) should be read, and certainly must have been read in the period, as a different kind of work than Robert Recorde's practical textbooks on arithmetic and astronomy, and also as different from Thomas Digges's learned Latin treatise on parallax and the supernova of 1572. It is also important to consider national differences. Because of the lingering influence of Scholasticism, the configurations of natural philosophy in France and England were quite different, so French texts cannot be cited, at least not uncritically, as evidence of the state of knowledge in England, even if they were being read in England.

Crucially, English Protestants went farther than any other European thinkers in expanding the number of phenomena considered to be purely natural and reducing the role of the supernatural.[59] Literary scholars have had a persistent tendency to misunderstand the early modern concepts of the occult and the supernatural, despite important work in the history of science that has clarified that early modern understanding of these terms was quite different from our modern sense of them.[60] We have misunderstood (and underestimated the sophistication of) early modern English ideas about the natural, the occult, and the supernatural and have attributed to early modern subjects a naiveté that is largely our own.[61]

A final idol, related to the preceding one, is what we might call the "New Age Renaissance," which places too much emphasis on early modern interest in Hermetic and other "magical" texts and in the elaboration of symbolic correspondences (Foucauldian "resemblance") that stitched nature together and rendered it meaningful.[62] The theories of Frances Yates, arguing for the importance of what she called "Hermeticism" in the early modern period, have been qualified and questioned by a number of historians of science, but literary scholars still sometimes cite her work as if it were unproblematic.[63] Certainly, as Grafton and Newman suggest, there are many texts in the period that evince such interests and call on such correspondences: images of microcosm and macrocosm are ubiquitous in early modern writing. However, Grafton, Newman, and others have suggested that alchemy and astrology were accepted on grounds that are more rational than modern scholars have tended to acknowledge.[64] And, as Mary Floyd-Wilson has argued against Foucauldian ideas about the visibility of analogies and similitudes, early modern belief in the existence of hidden "sympathies" and "antipathies" led to "natural philosophy's privileging of experimentation," and "fostered the development of Renaissance natural philosophy."[65] In the early parts of the sixteenth century, I contend, there was a strain of thought in England that conceived of a naturalistic discourse about the world that subordinated occult linkages to ordinary experience of nature and subjected them to rational scrutiny.

Interest in Hermetic texts, Paracelsan "signatures," and other secret disciplines proliferated, in England at least, toward the end of this period, as the intuitive certainty of Aristotelian natural philosophy was being rendered uncertain. English writers turned to these alternative forms of inquiry in part as a response to the loss of intuitive certainty and the accumulation of alarming natural phenomena (a new star, comets, an earthquake) that occurred in the 1570s, 1580s, and 1590s. Rather than constituting the dominant mode of premodern natural philosophy, "resemblance" emerged, in England at least, as a last-ditch effort to return the universe to a state of stability and intelligibility.

As I show, a number of writers in sixteenth-century England were aware of and understood the implications of Copernican astronomy and new theories of matter for the Aristotelian, Ptolemaic, and Galenic systems and the epistemology on which they were based. Recorde, Digges, Dee, Fulke, Thomas Harriot, and William Gilbert, among others, struggled to come to terms with ideas that seemed to threaten the stability and intelligibility of the universe, turning to various esoteric practices when direct observation of nature no longer seemed adequate to explain it. In these writers' textbooks, correspondence, marginalia, and treatises, we find common phrases and tropes that let us trace their shared preoccupations through the sixteenth century. These same phrases and tropes also appear in poems and plays written in England during the last twenty years of the century.

Although Spenser's *The Faerie Queene* has not generally been linked to the new science, it reflects many of the tensions and uncertainties that surfaced in England as Copernican theory and the supernova of 1572 called into question the stability of the universe and as the erosion of intuitive science threatened its intelligibility. In *Book V* and also in the *Mutabilitie Cantos*, Spenser tries to come to terms with new evidence that the whole universe was subject to change but ultimately turns away from what seems to have been an intolerable concept. Book II of *The Faerie Queene* explores the disjunction between characters' external appearance and demeanor and their internal humoral balance, linking his obsession with the legibility of surfaces to epistemological uncertainty connected with new ideas in medical writing. Here, too, Spenser pulls back from fully acknowledging the implications of the changes with which he grapples in his text.

While there is no evidence that Shakespeare shared Spenser's knowledge of or interest in contemporary astronomy, in his sonnets and in *King Lear* he alludes to new theories in mathematics and in terrestrial physics, facing squarely the existential and epistemological ruptures that Spenser skirted. The sonnets evince Shakespeare's interest in the new powers of abstraction afforded by Arabic numerals, especially by the concept of zero and its dizzying possibilities of multiplication—possibilities that seem generative in a positive way in the early sonnets addressed to the young man but that take on a negative cast when female sexuality becomes a factor in the later sonnets. The dark universe of *King Lear* is based at least in part on new concepts of nothingness, arising from the introduction of zero and also from the idea that matter is composed of atoms separated by void space, the fabric of the universe literally riddled with nothingness.

Theories of matter and motion also had political implications, and these are explored in Marlowe's *Tamburlaine* and Shakespeare's *Antony and Cleopatra. Tamburlaine* responds to some of the same controversies of the 1580s and 1590s that Spenser was concerned with in *The Faerie Queene*. Tamburlaine's conquests are imagined in meteorological and astronomical terms that link his transgressive ambition to a universe that seems newly subject to human agency and to change. Shakespeare's *Antony and Cleopatra* associates Egypt and Rome with different theories of matter, evincing what seems like nostalgia for the old vitalist system of elements and humors, even as it is conquered by a mechanistic world of hard, impervious, artificial surfaces.

I hope the title of this book will not mislead readers into thinking that premodern "intuitive" science was associated with some sort of ecological utopia before a fall into modern exploitation of nature.[66] Shakespeare may hint at the dangers of conceiving of the world as separable from the human body, but most of the writers I study here are eager to find new ways to control and exploit their environment. Charting this rupture probably will not provide a model for a "green" history of ideas.

This book may be modestly useful in calling attention to the problems we face today because science is so counterintuitive. Global warming and evolution happen on too large and long a scale to be perceived by the embodied mind. Brian Greene's book *The Hidden Reality: Parallel Universes and the Deep Laws of the Cosmos* begins with the observation that "if there was any doubt at the turn of the twentieth century, by the turn of the twenty-first, it was a foregone conclusion: when it comes to revealing the true nature of reality, common experience is deceptive."[67] Echoing a trope commonly found in early modern treatises like Digges's *Alae seu scalae mathematicae*, he explains that "mathematics" allows us to "peer behind everyday experiences" to reveal that "space, time, matter, and energy engage a behavioral repertoire unlike anything any of us have ever directly witnessed" (vii). Lewis Wolpert's *The Unnatural Nature of Science* confronts this problem head-on, seeking to understand contemporary hostility to science and widespread disbelief in scientific concepts such as evolution. He argues that "many of the misunderstandings about the nature of science might be corrected once it is realized just how 'unnatural' science is," unnatural because "scientific ideas are, with rare exceptions, counter-intuitive."[68]

Rather than viewing as irrational the right-wing rejection of evolution and global warming, it is probably more accurate to view acceptance of those arguments by someone like myself (who lacks advanced scientific training) as what Bruno Latour has described as "trust in the institution of science."[69] In both cases, a decision about whether or not to accept the word of scientists is driven, at least in part, by ideology. Wolpert acknowledges that "many people accept the ideas of science because they have been told that these ideas are true rather than because they understand them." He sees one solution to current popular distrust of science as a new kind of science education that would "compare common-sense ideas about the world with scientific views," arguing that children "need to appreciate just how different scientific thinking is and how much more natural were Aristotle's ideas as compared to those of Galileo and Newton."[70] Harvard physicist Eric Mazur has transformed his own approach to teaching physics, taking into account the intractability of intuitive ideas. To the extent that early modern literature reflects the loss of this "natural" way of thinking, it may help us come to terms with and think creatively about the problematic status of scientific knowledge in the twenty-first century.

CHAPTER 2

Aristotelian Naturalism and Its Discontents

My goal in this chapter is to determine the range of things that ordinary educated people in sixteenth-century England understood about the natural world and to trace the evolution of their thinking as radically new ideas, such as the heliocentric universe, the use of Hindu-Arabic numerals, the location of the supernova of 1572, and atomic theories of matter began to become known there. By "ordinary" and "educated," I mean literate people with a school or university education who had no particular expertise in mathematics or natural philosophy, people like the writers—Spenser, Shakespeare, and Donne—whose relation to changing ideas about the natural world are the subject of this study.[1] The orthodox answer to this question would be that their understanding of nature was mostly based on Aristotle, that it came from books and not from observation of nature itself, that it included an indiscriminate mix of actual material phenomena and superstitious magic, and that it was maintained only because the authority of ancient writers—preeminently Aristotle—could not be questioned. This understanding of nature was, as the story goes, quickly done away with once people (in the seventeenth century) turned away from reliance on the authority of books and began an empirical study of nature itself. It is also usually assumed that people in sixteenth-century England were largely ignorant of new ideas introduced by continental figures like Copernicus and Johannes Kepler. All of these ideas are actually at least partly wrong.[2]

In the first place, what it means to describe a body of knowledge as Aristotelian is such a complex issue that we should always imagine the term "Aristotle" as having invisible scare quotes around it.[3] First, we must sort out what constituted "Aristotelianisms" in sixteenth-century England and determine what aspects of that body of knowledge were basic and widely accepted. Second, the role of observation as opposed to textual authority in early modern concepts of the natural world is also complex, but there was more of a place for observation of a particular kind than has been widely acknowledged. Third, the relationship between observable phenomena—what we in the twenty-first century would consider actual "science"—and belief in supernatural or magical powers through astronomy, alchemy, Cabala, Hermeticism, and so on, was constantly being renegotiated in this period, but distinctions were made, albeit not in the same way we would make them.[4]

Finally, as I suggested in the introduction, the core of the "Aristotelian" understanding of the natural world coincided with intuitive science, folk theories about the natural world that arise to provide commonsense explanations of ordinary experience. This view of the world was supported by something deeper than reverence for authority, and writers who inveighed most vociferously against the authority of "Aristotle" and scholastic philosophy often continued to rely on many of its basic tenets.[5] After 1550, information about new theories and discoveries—such as the Copernican universe, changes in mathematical theory including the adoption of Hindu-Arabic numerals, the more accurate observation of parallax that identified a new star in the region of fixed stars, and ideas about atomism—were more widespread than many literary scholars realize. In addition, many of the new ideas about nature that replaced the old view—mechanistic atomism, an inertial theory of motion—were not subject to directly empirical demonstration in the seventeenth century. From the perspective of the late sixteenth century, the nascent scientific revolution might have seemed to involve a movement away from direct observation of natural phenomena, rather than toward it.

In my reading of treatises on the natural world written in sixteenth-century England, I have found it most useful to identify two basic approaches to understanding the universe that were inherited from antiquity and the middle ages.[6] Both were invested in describing a stable and intelligible universe, although the terms of its intelligibility were different in each case. One approach, which we think of as Aristotelian, was based on a belief that the natural world manifested itself to unaided human sense perception as it essentially was. This approach enshrined ordinary, commonsense, daily perceptual experience as necessarily the only way to access the truth about nature.[7] In practice, Aristotle seems to have read back from appearances to posit intelligible causes, although, of course, his system doesn't acknowledge this and asserts instead that the observed qualities reveal universal causes.[8] According to this approach, nature revealed its truths through ordinary experience but could not be made to do so through experiments since experiments forced nature to do unnatural things and therefore could not reveal any significant truths about it.[9]

The second and rival tradition focused on the aspects of nature that the Aristotelian tradition could not account for, the "occult," hidden, or "preternatural" phenomena that could not be explained in terms of the manifest qualities of the four elements.[10] James Bono traces this Renaissance tradition to the Neoplatonism of Marsilio Ficino, who sought to supplement "the merely manifest, often palpable, qualities (hot-cold, wet-dry)" with "a *scientia* of hidden powers and entities allowing access to the beneficial influences locked within the very essence of natural and celestial substances."[11] Bono suggests that Ficino's occultism "purports to 'read' sublunar and celestial nature by lifting the veils of manifest appearances, penetrating to the core of things themselves."[12] This tradition focused on the "secrets" of nature and found different ways to establish some body of knowledge or practice as providing access to those secrets: Platonism, alchemy, Hermeticism, Paracelsanism, Cabala. Geometry and optics were also threads in this tradition, which held that the universe was only intelligible to those who were adept in some specific practice.[13] Lorraine Daston has argued that "whereas Aristotelian natural philosophy had required only the most lightweight epistemological apparatus to study manifest properties and commonplace regularities, early modern preternatural philosophy needed heavier machinery to warrant knowledge of such elusive and ornery phenomena."[14] These approaches allowed for something closer to "experiment" in the modern sense because nature had to be forced to yield up its secrets. It is important to realize that *occult* in this period did not usually mean "supernatural" and that practices like astrology, alchemy, and natural magic spanned a range of natural and supernatural understandings of nature. In order to avoid confusion with the modern meaning of *occult*, I will refer to these practices as "secret."[15]

These two approaches to understanding the natural world coexisted, and a path to understanding the history of natural philosophy in sixteenth-century England lies in how these two traditions intermingled and were unsettled by, and tried to come to terms with, new discoveries in the last half of the century. The Copernican system, the supernova of 1572, advancements in mathematics, and speculations about atomism all challenged the terms of intelligibility and stability offered by both approaches. English writers who publicized continental discoveries in the later sixteenth century almost universally held a mix of orthodox Aristotelian views and beliefs that were part of the "secrets" tradition, although the balance between the two shifted in the course of the century. For these writers, familiarity with the idea that the truth about nature was hidden seems to have made the breakdown of the Aristotelian system more acceptable.

In order to trace the changing ideas about the natural world available to literate people in sixteenth-century England, I have relied on works published in the vernacular and in Latin, written for the most part by "experts" with specialized knowledge of mathematics, astronomy, and/or medicine, designed in most cases to convey up-to-date knowledge about nature to a wider public. These works reveal a concern with the relationship between sensible daily experience—the commonsense view of nature codified in popular Aristotelianism—and the increasingly counterintuitive theories being proposed by continental writers like Copernicus, Kepler, and Galileo. Early works (written before the 1550s) are fairly confident that ordinary experience of nature—aided, perhaps, by some instruction or rudimentary instruments—enables access to truth. Nature reveals its causes on the surface. As I trace in chapter 3, these writers begin to describe the "signs" and "tokens" that nature provides so that the educated subject can understand natural events.

In the middle of the century, a group of "mathematiques" connected with John Dee published a series of texts that made new ideas about mathematics, astronomy, and medicine available to a general public, introducing Arabic numerals, the algebra of the cossists, Copernican astronomy, and basic humoral diagnosis. These texts also believe in a legible universe but struggle to reconcile increasingly abstract and counterintuitive information with the data provided by ordinary sensory experience. Ideas based in Neoplatonism and other secrets traditions prove useful in explaining why some natural phenomena are not accurately perceptible by human sense. The supernova of 1572 was a watershed event, however. Members of the Dee circle (along with continental scholars like Tycho Brahe) were able for the first time to determine that the new star was located in the realm of the fixed stars, where no change was supposed to be possible. After this event, anxiety about the intelligibility and stability of the universe reached a crisis to which different writers reacted differently. By the end of the sixteenth century, long-held assumptions about the intelligibility of the universe in Aristotelian terms were radically questioned, with no clear theories as yet available to replace them. Some responded to these threats to the Aristotelian system with a hopeful pursuit of alternative secret knowledge; others believed that if the laws of the Aristotelian universe were in abeyance, the universe itself must be nearing its end. Gabriel Harvey and Francis Shakelton exemplify two divergent reactions to these emerging ideas by people without specialized training in mathematics or astronomy.

WHAT DID THEY MEAN BY "ARISTOTLE"?

"Aristotle" used as shorthand for early modern understanding of the natural world is both bigger and smaller than the corpus of Aristotle's writings on nature. It is generally understood that "Aristotle" denotes a larger body of work, including the writings of Aristotle that were extant in the sixteenth century (some of which are now understood to be pseudonymous), the corpus of Galen, and the astronomical writings in the Ptolemaic tradition, as well as medieval elaborations and reinterpretations of these bodies of work, especially those by Arabic writers like Avicenna and by Thomas Aquinas and his followers.[16]

However, only a fraction of this body of work comprised the basic understanding of nature in early modern England. Aristotle's theories of the natural world are complex and internally contradictory, articulated piecemeal across writings with very different topics and approaches.[17] Eckhard Kessler has usefully characterized the gap between Aristotle's "metaphysical" (Kessler's term) theories of nature and matter in the *Physics*, where he treats relatively abstract ideas about motion, change, and the four causes, and the more "naturalistic" account of the four elements and their perceptible qualities in *On Generation and Corruption* and the *Meteorology*.[18] These "naturalistic" writings are the source of theories that coincide with intuitive science and common sense. Not surprisingly, these are the ideas that are dominant in sixteenth-century English writing about nature.[19]

It is difficult to reconcile many details of the theoretical treatment of form, matter, and substance in the *Physics* with the more concrete theories of the four elements and the structure of the cosmos found in *On Generation and Corruption*, the *Meteorology*, and *On the Heavens*, although doing so was a major preoccupation of medieval philosophers who invented the concept of "substantial form" to resolve some of the most basic problems.[20] Aristotle's ideas about the teleology of motion and change—that all things necessarily move toward their destined end—are crucial to his metaphysical and naturalistic accounts. But his metaphysical theories about the relationship between form and matter are difficult to reconcile with the perceptible "qualities" that constitute the elements. Although the Aristotelian theory of substantial forms got a lot of attention in the seventeenth century as perhaps the most problematic aspect of his account of matter from the perspective of the new science, sixteenth-century English writers tended to emphasize the more concrete accounts of the four elements and to neglect the theories of form, matter, and causation unless they needed to deploy those metaphysical concepts for some specific purpose.[21]

Kessler argues that a number of Italian writers in the early sixteenth century began to turn away from the "metaphysical" account of nature found in the *Physics* and to focus instead on what Pietro Pomponazzi called *pura naturalia*, "dedicated to a purely empirical explanation of the sensible world of material reality, in the tradition of the Aristotelian treatise *On Generation and Corruption*."[22] Kessler acknowledges that those who taught philosophy at the university level continued to comment on and teach the theory of forms and causes found in the *Physics*; however, even these writers tended to distinguish between a *physica generalis* found in that text and a *physica particularis* found in the more concrete texts.[23] In England, while writers like John Case continued to provide commentary on the *Physics* and while university graduates were certainly familiar with hylomorphism and the four causes, accounts of the workings of nature tended to focus on the elements. Aristotle's distinction between substance and accident remained important in discussions of transubstantiation, and the distinction between material and formal causes on the one hand, efficient and final causes on the other, was useful in trying to determine whether phenomena like earthquakes and comets were naturally occurring events or warnings from God. Aristotle's multiple causes meant that they could be interpreted as both. Craig Martin has argued that the naturalistic *Meteorology* had broader influence than some of Aristotle's more abstract works because it was more accessible to those without university training, and more intuitive.[24] For the most part, English writers of the period, especially in vernacular writing, tended to focus on the "naturalistic" theories of the elements that they gleaned from cosmological treatises such as Joanes de Sacrobosco's mid-thirteenth-century *Sphaera Mundi*, Gregory Reisch's *Margarita philosophica* (1503), Marcellus Palengenius's *Zodiacus Vitae* (1531), and William Caxton's *Mirrour of the World* (1481).[25] Something akin to this "naturalistic" approach can also be found in orthodox Galenic medical texts like Thomas Elyot's *Castle of Health* and Robert Recorde's *Urinal of Physicke*, since Galen's humors correspond closely with Aristotle's elements.

Two popular pseudo-Aristotelian traditions reveal the bifurcation of natural philosophy into Aristotelian naturalism and secret "occultism." Ann Blair traces the popularity of the tradition of pseudo-Aristotelian *Problemata*, collections of questions and answers about the causes of "natural phenomena (especially relating to medicine, natural history, and meteorology)" in the early modern period.[26] She notes that "in its models and its general practice the genre is naturalistic, although it could be adapted to providential and moral conclusions."[27] Collections of *Problemata* were designed "to understand the causes of the regular phenomena of nature" and not the monstrous, marvelous, or occult.[28] They pose what Blair describes as "an apparent paradox" that might seem to trouble Aristotelian and

Galenic theory and then resolve it in terms of the qualities of the elements or humors that it initially seemed to violate.[29] She provides as an example this question: "Why is it that children, who are warm, are not fond of wine, while Scythians and courageous men, who are also warm, are fond of wine?"[30] The answer in this case is that children are "moist" as well as warm, whereas Scythians and courageous men are dry and therefore more drawn to liquid wine.

In this sense the *Problemata* were a counterpart to the also popular collections of pseudo-Aristotelian *Secreta,* which dealt with phenomena that were "manifestations of occult qualities, or events that occur unexpectedly or idiosyncratically as a result of insensible causes."[31] Gabriel Harvey, who often cites authors with opposing views as if there were no discrepancy, comments that "vulgar astrologers" of this day rely on "Albertus secrets, & Aristotles problems Inglished, their great Doctours & wonderfull Secreta secretorum."[32] Derived from an Arabic source but attributed to Aristotle, this encyclopedic work addressed political theory, medicine, astrology, alchemy, numerology, and magic and was, according to William Eamon, more widely known in the Middle Ages than Aristotle's canonical writings.[33] Roger Bacon was influenced by this tradition and its promise that secret knowledge of occult phenomena was available to those who knew how to pursue it.[34] Many vernacular secrets books were just collections of recipes, but some do articulate an alternative approach to reading nature: contrary to the orthodox Aristotelian position, the surface of nature doesn't directly reveal the truth about its causes, offering instead signs of hidden truth that only the adept can interpret.[35]

ARISTOTELIAN NATURALISM IN SIXTEENTH-CENTURY ENGLISH TEXTS

In early sixteenth-century England, the naturalistic Aristotelian tradition was dominant in popular writing about the natural world. The same basic account of the elements and the structure of the cosmos and the human body appears in a number of texts. This account is deeply essentialist and teleological, describing a universe in which the observable qualities of matter provide access to knowledge about its essential composition.[36] Although these theories may seem bizarre and obviously wrong to us now, Steven Shapin emphasizes that Aristotelian theories of matter and motion coincided with "common sense" and that "the resonance between traditional accounts of natural motion and the texture of human experience is evident."[37] John Rastell, in his "Interlude of the Four Elements," alludes to the importance of ordinary observation when he refers to "These four elementis here below, / Whose effectis dayly appere here at eye."[38] Below the moon, matter is composed of four elements, earth, water, air, and fire, which are constituted by their first qualities: hot, cold, wet, and dry. As Nature explains in Rastell's text, "The lower region callyd the elementall / Conteynynge these four elementis beloo: / The fyre, the ayre, the water, and yerth also," 35. This is quite different from our own understanding of "states" of matter, since the apparent state that a material substance takes is a direct reflection of the "qualities" of which it is composed. As Thomas Kuhn explains, in Aristotelian matter theory, "change occurs by changing qualities, not matter, by removing some qualities from some given matter and replacing them with others."[39]

As Kuhn further explains, motion is also a form of qualitative change in Aristotle's system, since position is "like wetness or hotness, a quality of the object."[40] The behavior and location of the elements were similarly determined by their nature; earth naturally seeks the center of the universe, so objects composed of earth tend to fall. Rastell's Nature instructs humankind that "The yerth as a poynt or center is situate / In the myddys of the worlde, with the water joyned" (36). This position is determined by its sensible qualities: "The yerth of it selfe is ponderous and hevy, / Colde and dry of his owne nature proper" (36). Again, it is important to emphasize that the appearance of matter told you crucial things about its essential nature, allowing you to explain why it behaved as it did—why things fell, rose, evaporated, grew, aged, died. Earth does appear cold, dry, and heavier than water or air; fire, similarly, is hot, dry, and lighter than earth, water, or air. What we would consider biological processes like growth, aging, and decay provided a model for thinking about all kinds of natural motion. Everything moved toward its destined end; the acorn toward an oak tree, a clod of earth toward the center of the universe. Artificially induced movement (violent motions) involved the application of a force that caused something to move away from its natural place, like a cannonball forced by an explosion to travel up into the air contrary to its nature.[41]

The properties of mixtures and transformation phenomena could be accounted for according to these theories, but they were counterintuitive. How could something with one set of qualities mix with something else yet still remain intact? How could water (cold, heavier than air) turn into steam (hot, lighter than air)? What happened to the original qualities during mixture or transmutation? Rastell provides an orthodox account of the solution to the problem of mixture that was worked out in the Middle Ages, but he gets somewhat tangled up in the explanation.

These elementis of them selfe so syngle be
Unto dyvers formys can not be devyded,

Yet they commyx togyder dayly ye see,
Whereof dyvers kyndes of thynges be ingenderyd,
Whiche thinges eftsoonys whan they be corruptyd,
Yche element I reduce to his furst estate,
So that nothynge can be utterly adnychelate. (35)

Despite the appeal to sense ("they commix togyder dayly ye see"), Nature must reveal her invisible workings ("I reduce to his furst estate") and resort to Latinate jargon ("adnychelate") to complete the explanation. It was around these problematic phenomena that more esoteric theories like alchemy, Paracelsanism, and, ultimately, atomism tended to grow up since they were not easily explained in orthodox Aristotelian terms.[42]

In this basic Aristotelian system, forces or movements that could not be explained with reference to the sensible qualities were considered to be "occult" or, literally in Latin, "hidden."[43] A common example of an occult phenomenon in early modern writing was the force of magnetism. That some rocks were drawn to each other and that a magnetized needle pointed north were facts not intelligible in terms of perceptible qualities of matter: the rocks were made of earth, but the coldness, dryness, and weight of earth could not explain magnetic attraction. As William Eamon explains, Aristotelian *scientia* consisted of explanations based on "the manifest qualities of the four terrestrial elements," while occult qualities, like magnetism, involved "events that occur unexpectedly or idiosyncratically as a result of insensible causes."[44] In Aristotelian theory, therefore, "occult" phenomena were considered to be inexplicable and unintelligible, since they could not be perceived by the senses or explained in terms of the four elements and their qualities.[45] Occult phenomena ultimately came to be regarded as faultlines that encouraged speculation about the "secrets" of nature, but it is important to understand that many writers in the period believed these phenomena to have natural explanations and didn't necessarily represent supernatural phenomena.[46] Therefore, so-called occult sciences like alchemy and astronomy and even natural magic usually dealt in terms of natural, rather than supernatural, explanations.[47]

In the Aristotelian system, a wide variety of observed terrestrial and atmospheric phenomena came under the rubric of meteorology. As S. K. Heninger explains in his very useful *Handbook of Renaissance Meteorology*, the term *meteor* "included all atmospheric phenomena—that is, all natural processes that occurred in the region of Air: clouds, dew, winds, lightning, comets, rainbows, and associated weather processes."[48] Earthquakes were also meteorological phenomena because they were believed to be caused by "exhalations" trapped inside the Earth.

Meteorological events occurred when the sun (or other heavenly bodies) drew up "exhalations"—hot dry evaporation from the earth—or "vapours"—cold, moist evaporation from water—into the upper regions of the sublunar atmosphere.[49] Up in the atmosphere they were acted upon by changes in temperature: "From the dry exhalation came thunder and lightening, the winds, earthquakes, comets, and other fiery impressions; from the moist vapor developed clouds, rain, snow, hail, mist, dew, and frost."[50] Comets occurred when exhalations approached the fiery sphere and were ignited.

Aristotelian science provided the accepted natural explanations for these events, but because God functioned as the efficient and final cause of all natural phenomena, they could also be construed as omens or warnings.[51] Leonard Digges, writing in his A *prognostication everlasting* in 1555, reflects the kinship of comets and earthquakes as meteorological events when he notes that "Comets signifie corruption of the ayre. They are signes of Earthquakes, of wars, changing of kingdoms, great dearth of Corne, yea a common death of man and beast" (6r). Although he goes beyond a naturalistic account when he links meteorological phenomena to the prediction of dire events, he qualifies these predictions by citing Pontanus in Latin identifying comets as predicting wars, adding references in the margin to Cardanus and Mizaldus's "De Cometarum prodigiis" (6r). Digges provides Latin citations only when making claims about the supernatural significance of natural events. He notes similarly about earthquakes: "Plentie of winds, succed into holes, cones, or caves of the earth, which absent from above the earth causeth quietnesse: the violent brusting out of them (the earth closed againe) is the Earthquake: signum est futurorum bellorum" (13v). Elsewhere, he describes how thunder can predict weather, "Thunderes in the morning, signifie wind: About noone, raine," but questions other kinds of prediction: "Some write (their ground I see not) that Sundayes thunder, should bring the death of learned men, Judges, and others" (6v).[52]

Above the moon, things were not composed of the four elements but of a different, unchanging substance, the ether or "quintessence," which held concentric crystalline spheres that carried the heavenly bodies around the earth. Rastell's Nature explains that "The great world beholde, lo, devydyd wondersly/In to two regyons, wherof on I call/The etheriall region with the hevyns hye, /Conteynynge the planettys, sterris and speris all" (35). Robert Recorde, writing in his astronomical treatise *The Castle of Knowledge* (1556), describes how the earthly region of the elements located below the sphere of the moon differs from the "rest of the worlde, which hath no suche alteration nor corruption, whiche parte is above all the foure elementes, and compasseth them about, and is called the Skie, or

Welkin, and also the Heavens" (7). The elements on Earth would have remained static, arranged concentrically according to their natures (earth, surrounded by water, surrounded by air, surrounded by fire), except they have been disturbed by forces emanating from the heavenly bodies.

The unchanging part of the heavens consisted of a series of concentric spheres, one for the sun, the moon, and each of the known planets, enclosed by the outermost sphere of the fixed stars. Ptolemaic astronomy added to this structure a complicated system of circles and epicycles to accommodate the observed movements of the planets to a geocentric universe.[53] Two observed motions were particularly difficult to account for according to the Ptolemaic model. First, because planetary orbits around the sun are completed at different rates, the usual observed planetary movement across the sky from east to west periodically appears to stop and move westward in retrograde motion, as the orbit of the earth overtakes or is overtaken by that of the planet.[54] Ptolemaic astronomy accounted for retrograde planetary motion with a complicated system of epicycles and deferents, which were necessary to explain what was actually an optical illusion produced by the motion of the earth.[55] As Thomas Kuhn explains, this theory of epicycles was "an astounding achievement," but "it never quite worked" and was in part what motivated Copernicus to develop an alternative explanation.[56] Second, because of a wobble in the rotation of the earth, the fixed stars appear to move in a very long (26,000 year) cycle, with periodic retrograde motions called "precession" or "trepidation."[57] Ptolemaic astronomers posited the existence of a ninth and a tenth sphere to explain this movement, but this theory was always troubling. The fixity of the stars secured the stability of the universe, so it was an issue if they were seen to move, and it was also a problem that the hypothetical ninth and tenth spheres held no heavenly bodies to make them "visible" like the other eight spheres.

Rastell's Nature avoids going into detail on these complicated planetary motions: "Every sterre and spere in straunge maner/Upon his owne poles movyth dyversly, /Whiche now to declare were to longe to tary" (37). Similarly, when Recorde's Scholar asks the Master to explain "the progression, retrogradation, and station of the Planetes," the Master refuses because, he says, the topic is too "obscure" for the Scholar's elementary grasp of astronomy (279).

The sublunary world was mortal, chaotic, always changing—but it was stabilized by the sphere of the fixed stars. And it was rendered intelligible by the manifest natures of the elements and by our ability to read what was happening in the heavens. It was clear that the sun and moon have powerful effects on the elemental world—the sun causes plants to grow, the moon influences tides—so it made sense by extrapolation to assume that the planets and stars have similarly powerful if not directly visible effects on the earth. If we are tempted to ridicule this early modern belief that the universe was full of powerful invisible forces, we should recall that gravity, electricity, and radiation are all examples of such forces.

Reading the heavens is a place where Aristotelian meteorology again blends—seamlessly and gradually—into more esoteric practices like astrology. Astrology was divided into "natural astrology," the study of the position of the heavenly bodies and the nature of the rays that they transmit (which almost everyone accepted as a valid pursuit), and "judicial astrology," which involved making specific predictions about human events based on the position of the heavenly bodies (which was widely questioned, partly on religious grounds).[58] A treatise like Leonard Digges's A *prognostication everlasting* (1555) begins with basic meteorology and explains how to predict weather by reading signs in the sky: "The Sunne in the Horizon or rising cleare and bright, sheweth a pleasant day: but thinly overcast with a cloude betokeneth foule weather" (5r). In addition to commonsense naturalistic observations such as this one, the book also includes how to predict weather based on the natural qualities of astrological signs: "if she [the moon] be in hote and drie signes, as Aries, Leo, Sagittarius, in winter a good token of faire weather" (5v). He explains the use of astrology to determine proper medical treatment because of the power of the moon over water: if the moon "dispersing her vertue, She filleth all places with moisture," it must also be true that "at the full the Moone lodeth all bodies with humors" (20r). As we saw above, he sometimes cites other authors who claim that meteorological events can predict war or other human calamities, but he never directly makes those claims himself.

John Dee's seemingly bizarre theories of astrology and communication with angels are based on this idea that all heavenly bodies send out "species" or rays. Rather than espousing an atomic theory of matter, Dee believed that many natural phenomena could be explained by the action of these invisible rays. In his *Propaedeumata Aphoristica* of 1558, Dee, writing in the disjointed aphoristic style associated with the "secrets" tradition, lays out a theory that sees the key to deeper understanding of the universe as a systematic quantification of the rays emitted by heavenly bodies. As J. L. Heilbron explains, Dee based his theories of planetary species on the behavior of "visible light," which meant that he could employ "all geometrical optics then known" as a means of calculating "the relative strengths of astrological radiations in different planetary configurations." Heilbron also describes how Dee's interest in optics extended to experiments with concave mirrors to bend and focus rays of light, a practice that he based on work by Roger Bacon and called "Catoptrics."[59] Deborah Harkness argues that the crystal "showstone" used by Dee's scryer to receive messages from Angels was, in Dee's mind, based in "the physics and metaphysics of light" outlined in the *Propaedeumata*.[60] Harkness sees Dee's angel conversations as involving a "tangle of religious, magical, and natural philosophical ideas" but emphasizes that his use of the showstone to focus emanations from angels was plausibly "the culmination of his natural philosophy."[61]

GALENIC NATURALISM

Medical practitioners in early sixteenth-century England ranged from dispensers of folk remedies to trained apothecaries and surgeons to university-educated physicians.[62] Educated Londoners would have been familiar with physicians, who were concentrated in London and whose practice was based on the system of Galenic medicine, inherited from antiquity, with its theory of humors and temperament or complexion. Galenic medicine differed from Aristotelian natural philosophy in some significant ways, but its theory of humors was strongly influenced by Aristotle's naturalistic accounts of the elements.[63] Like Aristotelian physics, Galenic medicine seems self-evidently wrong to us now. Gail Paster, who has provided one of the most compelling and sympathetic contemporary accounts of Galenic humoralism, describes it as "the residuum of error."[64]

Like Aristotle's theories of matter, however, humoralism was based on reading back from observed phenomena—the symptoms of disease—to posit causes that were qualitatively linked to what was observed. As Nancy Siraisi notes, a person's "complexion" or balance of humours "was supposed to be a perceptible quality that could be discerned by touch."[65] To this extent, then, Galenic medicine shared the Aristotelian epistemology of the manifest. Symptoms of disease tend to involve alterations in temperature of the body and alterations in the appearance and behavior of its fluids: fever, sweating, blowing your nose, sneezing, coughing, vomiting, diarrhea, and bleeding. It was a rational conclusion, therefore, to link the causes of disease with the balance and regulation of fluids within the body, the four humors (blood, phlegm, black bile, and yellow bile), which shared the same qualities of warm, cold, moist, and dry as the elements and were linked to them.

A treatise like Thomas Elyot's *The Castel of Health* (1541) explains how a state of health results from a proper balance of the humors within the body and how disease results from an imbalance caused by external forces such as food, air, and climate. Elyot teaches that "complexyon" is "a combynation of two dyverse qualities of the iiii elementes in one body," noting, for example, that "the body, where heate and moisture have soverayntie is called Sanguyne" (2r). Disease is caused by "distemperaunce," which "bryngeth sycknesse or griefe into the membre" (3r).

In this system, disease can be caused by, for example, corrupted air, which is in turn caused by factors ranging from "influences of sundry sterres," to "great standyng waters never refreshed" (12r).

Through the sixteenth century in England, Galenic medicine remained the dominant learned medical tradition. Vesalius's groundbreaking anatomical treatise, *De Humani Corporis Fabrica* (1543), was known in England, and a few dissections were allowed there each year beginning in the early sixteenth century, but the study of human anatomy through dissection did not significantly disrupt Galenic medicine until later in the seventeenth century.[66] Although dissection allowed a view inside the human body, the bodily interior of a living sick person remained hidden from the physician, who could only diagnose disease based on manifest humoral symptoms.

We can see some signs of threat to the Galenic system in the sixteenth-century reaction to "new" diseases like syphilis and bubonic plague, which were not known to Galen and which seemed to be spread by some means of contagion (rather than arising from an imbalance of humors). As Siraisi notes, "the issue of the possible existence of diseases inexplicable in terms of complexion emerged more clearly and was the subject of intensive discussion during the sixteenth century."[67] The causes of contagious disease were literally occult, since the means of transmission could not be observed, and medical writers in the period argued over how best to explain these diseases.[68] The "miasma" theory that corrupted air spread disease provided a way to render contagion perceptible, since the bad smell could be perceived even if the means of transmission could not be seen. This explanation could not explain every instance of contagion, though, so speculation about the invisible means of transmission sometimes lead to anti-Aristotelian theories. Some medieval Arabic writers had posited the idea that disease was spread by invisible bodies, and the Italian humanist Girolamo Fracastoro, influenced by Lucretian atomism and Lucretius's account of tiny "semina morbi," developed a similar theory in his 1546 poem on syphilis.[69] The German anti-Aristotelian Paracelsus similarly linked a sort of germ theory of disease with, as Siraisi puts it, a larger goal of repudiating "the entire tradition of academic medicine," replacing it with "a different system based on folk medicine and craft wisdom, alchemy, and occult practices."[70] Paracelsus astutely recognized that the Galenic humors and qualities (hot, cold, moist, dry) were not causes of disease but rather "after-effects, products of the disease, rather than primary factors in its causation."[71] As Richard Bostocke explains in the first comprehensive Paracelsan medical text in English, "humours and qualities" are "conditions, signs, tokens, and as it were onely flowers and colours of diseases, and not the very matter, cause, substance, or nature of the disease."[72] He believed, instead, that diseases were caused by minerals spread through the air as tiny "seeds" of disease.[73]

Building on this alternative theory of disease, Paracelsus replaced the easily read natural world of Aristotle and Galen with a semisecret network of occult resemblance and "signatures." James Bono explains Paracelsus's belief that "Aristotelians, Galenists, and other postlapsarian students of nature" sought, mistakenly, to read nature by observing its manifest qualities. Paracelsus believed, instead, that "the properties, powers, or natures that give rise to the activity of things in nature are occult, i.e., hidden from our direct apprehension," but that "nature has left evident 'traces' of such animating properties in the visible texture of things themselves."[74] Paracelsus believed that these "signatures" could be read by those who became "adept at reading signs," that is, those who learned and applied the Paracelsan system.[75] In Bostocke's adaptation of Paracelsan theory, the chemical physician uses distillation to "purge and clense" substances of "their outward forms leaving only their Arcana, inward and secret things remayning for the physition to judge upon" (D6v). Through visible signs, various plants and minerals reveal which diseases or parts of the body they can cure. So, since cancer in the body corresponds to deposits of arsenic in the earth, arsenic cures cancer; since the thistle plant has leaves that prick like needles, it can cure "internal stitches."[76]

Although Foucault, among others, takes the Paracelsan system as representative of a premodern approach to the natural world, it in fact marked a sharp departure from Aristotelian physical theory, and it is more accurate to see it in the context of a "secrets" tradition that coexisted with and countered orthodox Aristotelianism. Jonathan Sawday describes Paracelsus as "central to the defence of the old intellectual order," missing its significant differences from Aristotelian orthodoxy but rightly noting its divergence from the new science.[77] The Aristotelian and Paracelsan traditions offered a method of understanding concrete things in nature through visible signs, and both differed from new approaches to nature that posited a reality that deviated sharply from its appearances and that approached natural processes through abstraction from concrete events.[78]

MATHEMATICAL ESSENTIALISM

Mathematics in this period also inherited from classical antiquity an essentialist tradition that was supplanted by more abstract ideas as Arabic numerals, computation on paper rather than with an abacus, and early forms of algebra were developed. However, the trajectory of mathematical thought from antiquity to the sixteenth century was not straightforward. Historians of science divide ancient mathematics into two traditions or "cultures": a practical tradition based on arithmetic calculation with pebbles or counters and geometric measuring with ropes, and an elite theoretical tradition reflected in the writing of Plato, Euclid, and others.[79] The theoretical tradition of ancient mathematics differed from astronomy and medicine in being more strongly influenced by Platonic, rather than Aristotelian ideas. Plato regarded numbers and geometrical concepts as abstractions.[80] Aristotle viewed "mathematical objects" as "a class of ideas mediating between the sensible objects themselves and the essence of objects," but he thought that they "had no reality apart from visible and tangible things."[81] As Alfred Crosby has argued, Aristotle did not particularly value mathematics because he felt that qualitative analysis was more useful than a focus on the quantitative, a situation that persisted in Europe until the sixteenth century.[82]

The practical arithmetic tradition of reckoning with counters was for many years transmitted orally rather than in writing, but it had profound effects on the shape of ancient mathematics, and it survived as the dominant form of arithmetic calculation in Europe until the sixteenth century.[83] When we think of an abacus we tend to think of the eastern version that used beads moved along wires, but Europe inherited from classical antiquity a system that involved placing pebbles or counters on a board marked with lines.[84] Reviel Netz has argued that the practice of calculating with counters was based on embodied intuition in the form of two fundamental human physical and cognitive capacities, the opposable thumb and "subitization," the ability to immediately perceive the number of objects contained in a small cluster of four or five objects.[85] He suggests that ancient theoretical mathematics focused on geometry and marginalized the numerical because numeration was so strongly associated with the practical (and socially inferior) use of counters. It may be that the several well-known conceptual limitations of ancient arithmetic also derived in part from the counter tradition, since these limitations arise from a tradition of counting. Ancient arithmetic could not deal with irrational numbers, and, as a result, geometry (which dealt with continuous magnitude, which would necessarily involve irrational numbers) was separated from arithmetic (which deals with multitude and numbers). Ancient mathematics did not have a concept of zero, and one was not considered to be a number because it defines the kind of thing that is being counted: one sheep is not "one," it is a sheep.[86] Only more than one constitutes a multiplicity of things and therefore a number: two sheep is two. One was also thought to be indivisible, since in counting you enumerate discrete, whole objects, and sixteenth-century English mathematicians inherited a mathematics that in theory dealt only with whole numbers and did not believe that fractions could exist. As Robert Recorde explains in his treatise on elementary algebra, *The Whetstone of Witte* (1557), only whole numbers "of Euclid, Boetius, and other good writers are called nombres." Fractions are "broken numbers" because "all nombers doe consiste of a multitude of unities: and ever proper fraction is lesse than an unitie, and therefore can not fractions exactly be called numbers."[87]

Morris Kline has argued that Greek mathematicians could not accept the existence of irrational numbers because "the intuitive basis of the idealization was not clear."[88] That is, despite the importance of abstraction in Pythagorean and Platonic arithmetic, irrational numbers were not acceptable because they did not correspond with intuitive concepts of the world. Greek mathematicians also "failed to comprehend the infinitely large, the infinitely small, and infinite processes," bound to an Aristotelian notion that "only as objects are delimited and distinct do they have a nature."[89]

Although Diophantus and other ancient thinkers developed mathematics in directions that pushed against these limitations, their work was not widely known until much later, when algebra began to be developed in the late sixteenth and early seventeenth centuries.[90] The Romans focused on practical engineering and did not make advancements in mathematics, so the math inherited from antiquity into the Middle Ages and ultimately in sixteenth-century England was not particularly sophisticated. Calculation was done on a board with counters and consisted of addition, subtraction, multiplication, and division using integers. Fractions were rarely used, and irrational numbers did not occur.[91]

Ordinary educated people probably did not know much math, except for merchants who began to attend special reckoning schools to learn basic mathematical and bookkeeping skills.[92] Those with a university education would have been familiar with Boethius's theoretical *Arithmetica*, which was part of the curriculum.[93] They learned a version of ancient arithmetic that was still by nature tied to its role in enumerating objects and influenced by the continuing use of abacus-like devices for calculation. Although mathematicians like Francoise Viète in France and Raphael Bombelli and Niccolo Tartaglia in Italy were developing a more sophisticated algebraic method in the sixteenth century, and although Thomas Harriot in England knew the work of Viète and developed some similarly advanced algebraic concepts, these ideas were not generally known in England until the seventeenth century.

Emphasis on the nature of things being counted meant that different kinds of things were thought to be incommensurable and could not be included in the same equation. Adding sheep and goats doesn't tell you anything unless you're willing to move to a level of abstraction (farm animal) beyond the identity of the particular beings you're counting, a move that ancient mathematicians were unwilling to make in theory. As Recorde explains in *The Grounde of Artes* (1540) "in going about to adde together the summes of sundrie thinges, which you ought not to do, except you seeke only the number of them and care not for the thinges" (50). This incommensurability rule proved to be a barrier to progress in the physics of motion until, in the fifteenth and sixteenth centuries, people like Galileo began to include distance and time in the same equation.[94]

In sixteenth-century England, then, the most significant mathematical advance involved the introduction of Hindu-Arabic numerals, including the concept of zero, duodecimal fractions, place value, and the practice of reckoning with a pen rather than counters.[95] The efficiency of computation made possible by the adoption of Hindu-Arabic numerals was a necessary precondition for the later development of algebra.[96] The fifteenth and sixteenth centuries saw European advances in mathematical knowledge and more widespread dissemination of this new knowledge, in part because increased trade required more complex accounting practices than could be carried out using Roman numerals and the abacus. Double-entry bookkeeping was developed, and merchants began to use a kind of basic algebra called the "Rule of three" or "golden rule" to figure out unknowns involving more than one kind of thing in the same equation.[97] In England, those with advanced mathematical knowledge, like John Dee, Thomas Digges, Robert Recorde, and Thomas Harriot, were familiar with techniques of reckoning with a pen, Euclidean geometry, the "cossic" algebra developed by merchants, and the basic trigonometry used to determine the location of heavenly bodies. These men were at the forefront of importing new ideas from abroad, largely because they were in a better position to understand the mathematical complexities that had accumulated as generations of observers attempted to "save the phenomena" of the Ptolemaic system, and therefore they could appreciate the simplification offered by Copernicus.[98] These mathematicians all to some extent held Neoplatonic ideas about geometry that allied some of their mathematical thought with the "secrets" tradition. A broader educated public in England did not begin to understand the more complex algebra and trigonometry that these writers learned from continental mathematicians until after 1610.[99] Recorde's influential treatises on arithmetic and basic "cossic" algebra reveal the extent to which the introduction of Arabic numerals, place value, and zero marked the most significant conceptual change in sixteenth-century English mathematics, shifting away from concrete manipulation of counters to a more abstract numerical representational system.[100]

NEW IDEAS THREATEN ARISTOTELIAN NATURALISM

Beginning in the middle of the sixteenth century, a series of discoveries and new technologies began to threaten the accepted view of a stable and intelligible universe whose surface appearance provided a reliable guide to its essential nature. These discoveries also engaged and altered the traditions that approached nature as revealing hidden causes to those in possession of secret knowledge. Most important innovations and discoveries occurred elsewhere in Europe, but there were two circles of English thinkers who were in touch with natural philosophers and mathematicians abroad and who wrote works that popularized their new ideas for a wider English public: the group of "mathematiques" just discussed, including Dee, Thomas Digges, and Recorde, and another group that formed slightly later around Henry Percy, Earl of Northumberland, and included the mathematician Thomas Harriot.[101] The works of the Dee group were more widely known because they relied on print publication; however, there is evidence that Harriot's ideas about atomism, if not his more advanced work in algebra, were familiar to people who knew him and his circle. Both groups espoused theories of the natural world that incorporated orthodox Aristotelian naturalism, as well as ideas derived from "secret" traditions.

Important publications by these writers include Robert Recorde's *Castle of Knowledge* (1556), which contains the first reference to Copernican theory published in English, and Thomas Digges's *A Perfit Description of the Celestiall Orbes* (1576), which provides an extended account and endorsement of the Copernican universe.[102] Recorde also published two important mathematical works, *The Grounde of Artes* (1543), which introduced Arabic numerals, zero, and place value, and *The Whetstone of Witte* (1557), which taught basic algebra. Digges published *Alae seu scalae mathematicae* in 1573, using parallax to demonstrate that the supernova of 1572 was located in the realm of the fixed stars. Dee also wrote a treatise on the supernova and published a "Mathematicall Preface" to Henry Billingsley's edition of *The Elements of Geometrie of Euclid of Megara* (1570), providing a wide-ranging and ambitious description of what mathematics was capable of. Dee's *Propaedeumata Aphoristica* (1558 and 1568) attempted to set out a method for quantifying the rays or "species" emitted by celestial bodies as a means of providing a scientific basis for astrology; however, this work straddles the line between naturalistic science and the secrets tradition.[103] Harriot corresponded with Kepler and left manuscript pages recording advanced speculations on algebra, geometry, astronomy, optics, and atomism.[104] Harriot and his circle had a reputation for atheism and experimentation with magic, and his speculations, too, were informed by ideas derived from natural magic and from the works of Roger Bacon.[105]

Robert Recorde's *Castle of Knowledge*, published in 1556, was "the first comprehensive, original treatise on the elements of astronomy to be printed in English."[106] Taking the form of a dialogue between a Master and a Student, it covers the basics of Ptolemaic astronomy, including instructions on how to construct and understand an armillary sphere, a 3-D model of the Ptolemaic universe. Francis R. Johnson argues that treatises on the use of instruments like the armillary sphere or astrolabe must assume a geocentric universe because they need to represent the heavens as seen from the perspective of the observer on Earth.[107] It is not necessarily a contradiction in terms, then, when near the end of the book the Master seems to imply that he is open to Copernican theory: "Copernicus a man of greate learninge, of muche experience, and of wondrefull diligence in observation, hathe renewed the opinion of Aristarcus Samius, and affirmeth that the earthe not only moveth circularlye about his owne centre, but also may be, yea and is, continually out of the precise centre of the world 38 hundreth thousand miles" (Recorde, 165). The Master concludes that this is too complicated to get into in an introductory treatise: "but bicause the understanding of that controversy dependeth of profounder knowledg then in this Introduction may be uttered conveniently, I will let it passe tyll some other time." The Student replies indignantly that he does not want to "heare such vaine phantasies." The Master says that the student doesn't yet know enough to evaluate the evidence, but when he is more knowledgeable the Master will "so declare his supposition, that you shall not only wonder to hear it, but also peradventure be as earnest then to credite it, as you are now to condemne it" (165). English readers would thus have been introduced to the Copernican model of the universe as early as 1556.

Thomas Digges provided a more detailed account of the Copernican universe in an appendix to a new edition of his father Leonard Digges's *A prognostication everlasting* in 1576, entitled *A Perfit Description of the Caelestiall Orbes according to the most aunciente doctrine of the Pythagoreans, latelye revived by Copernicus and by Geometricall Demonstrations approved*. Digges essentially provides an English translation of some opening sections of the *De revolutionibus* and strongly endorses the heliocentric universe.[108] Johnson notes the popularity of this work, which was reprinted six more times between the 1576 edition and 1605.[109] Digges explains that his father's almanac had assumed a mistaken "Modill of the world, and situation of Spheres Coelestiall and elementare according to the doctrine of Ptolomie, whereunto all universities (led thereunto chiefly by the authoritie of Aristotle) sithens have consented" (M1r). Although in other places Digges notes the coincidence of the Ptolemaic universe with the data of sense, he here makes the common accusation that continued belief in a heliocentric universe is based on reverence for the authority of Aristotle. Digges then reveals that he intends to provide readers with knowledge of "a new Theorick or Modill of the world, shewing that the Earth resteth not in the Center of the whole world" but is instead "caryed yearely round about the Sunne" (M1r). Familiarity with Platonism and, perhaps, the lack of a university education grounded on Aristotle, seems to make it easier for Digges to accept a heliocentric universe; he says that the sun "like a king in the middest of all raygneth and giveth lawes of motion to the rest, sphaerically dispersing his glorious beames of light throughout this sacred Coelestiall Temple" (M1r).[110] Digges goes on to debunk the Aristotelian argument that the Earth must lie motionless at the center of the universe because of the intrinsic and manifest heaviness of the element earth, and he offers instead secret knowledge, teaching people how "with Geometricall eyes to beehoulde the secrete perfection of *Copernicus Theoricke*" (N4r).[111] Digges did not, however, reproduce Copernicus's mathematical tables and trigonometric proofs, perhaps because he was aware that his audience did not possess the necessary mathematical knowledge to understand them.

John Dee, on the other hand, was aware of the Copernican hypothesis as early as Digges and like him appreciated that it provided a convincing mathematical account of the universe. However, he seems to have been reluctant to accept it as a physical model of the universe.[112] Stephen Johnston speculates that this was because Dee was more grounded than Digges in Aristotelian science and had a stronger commitment to a philosophical, rather than just mathematical, vision of the universe.[113] It may also be that Dee's complex astrological theories were dependent on a geocentric universe, so he was reluctant to let it go for that reason.

Digges's belief in the "secret perfection" of a geometrical universe may be what allowed him to retain a belief that the universe is stable and intelligible even in the face of the most devastating discovery of the late sixteenth century: the realization that a new star that appeared in the region of Cassiopeia in 1572 was located in the supposedly unchanging region of fixed stars. Aristotelian cosmology was dependent on the idea that the mutable "elementary" world below the moon was stabilized by the immutability of the fixed stars located beyond it. James Lattis explains the importance of this new star to the erosion of belief in the Ptolemaic universe: "If one considers comets to be within the terrestrial atmosphere, then the only new phenomenon to present itself in the history of astronomy before 1572 was the apparent change in the rate of precession of the equinoxes. But the effects first of Thabit ibn Qurra and then of the Alfonsine astronomers had succeeded in accommodating variable precession, or trepidation, with Ptolemaic cosmology through the introduction of invisible superfirmamental spheres to produce motions of the 8th sphere as required."[114] Francis Johnson argues that "the research and discoveries in connection with this famous new star did more than anything else to bring about the downfall of Aristotelian cosmology."[115]

In his *Propaedumata*, published in 1568 just before the appearance of the new star, Dee reflects on the importance of the immutable fixity of the stars: "That the mutual spaces among the fixed stars have never been altered in the whole eternity of time shows that the stars are very much superior even to those things in the elemental universe that strongly retain an unvarying condition in their situations." He goes on to say, "If this were not so [if the fixed stars were not immutable] no particle would be preserved naturally even for a single day."[116]

When the supernova appeared in November of 1572, it was readily visible with the naked eye and caused much consternation. Elizabeth I and William Cecil consulted mathematicians and astrologers in an attempt to understand what it was and what it meant. New scientific instruments and a new awareness of the importance of accurate observation made it possible for astronomers like Digges, Tycho Brahe, and Dee to determine that the new star had no visible parallax, and therefore could not be located below the moon.[117] This fact could be variously interpreted. It could suggest that the Ptolemaic/Aristotelian system was incorrect, or it could suggest that the star had been placed in the heavens, contrary to the laws of the Aristotelian cosmos, by God as an omen or portent, much as God had suspended natural law when he created the star to guide the Magi to Christ's birth and when he caused an eclipse at Christ's death. Another possibility was that it signified that the universe was on its last legs, becoming too old to function according to the laws it had previously followed.[118] Digges's treatise on the star and its location, *Alae seu scalae mathematicae* [Mathematical wings or stairs] (1573), argued forcefully that the star was not a comet but was indeed a new star located in the supposedly unchanging realm of the fixed stars. Digges is noncommittal about the significance of the star's location, suggesting that it might provide evidence in support of the Copernican system or that it might be a supernatural event like the star that led the Magi. Digges's willingness to debunk Aristotle in his 1576 work on Copernicus may rely in part on his observations of the new star.

But as Johnson and Deborah Harkness acknowledge, many saw the new star as a sign that the cosmos was in decline and perhaps close to an end. Harkness suggests that the supernova, an earthquake that occurred in England in 1580, and the prospect of a rare astrological conjunction of Saturn and Jupiter in 1583 led a number of people, including probably Dee, to believe that the world was becoming unstable and perhaps approaching its end.[119] She suggests that Dee turned to conversations with angels as a way to "address the difficulties associated with practicing natural philosophy at a time when the natural world seemed unreliable and mutable."[120] A number of writers in the period alluded to the new star or the 1580 earthquake and struggled to understand whether they were natural or supernatural events. They were clearly signs of something, but of what? The end of the Aristotelian understanding of the world or of the world itself?

TWO "ORDINARY" READERS

Gabriel Harvey and Francis Shakelton have little in common other than that both responded in writing to some of the new ideas published by Digges, Recorde, and Dee, and both addressed the question of how to determine whether events like the 1580 earthquake and the supernova were natural or supernatural in origin. Both had a clear basic grounding in Aristotelian naturalism and considered new ideas in that context. Nothing much is known about Shakelton except that he is identified as "A Minister and preacher of the worde of God" on the title page of his 1580 treatise entitled *A Blazyng Starre or burnyng Beacon, seene the 10. Of October laste (and yet continewyng) set on fire by Gods providence, to call all sinners to earnest & speedie repentance.* Literary scholars know Harvey mostly for his correspondence with his friend Edmund Spenser about quantitative meter and for his involvement in an exchange of insults with Thomas Nashe and Robert Greene on the fringes of the Marprelate controversy.[121] He is perhaps not exactly an "ordinary" reader of these works because his brothers seem to have had specialized training in natural philosophy. Richard Harvey published an infamous astrological treatise in 1583 called *An Astrological Discourse upon the Conjunction of Saturn and Jupiter* which predicted that the conjunction of two major planets on April 28 and other astrological events in 1583 would mark the end of the world. Most astrological writers knew to avoid such specific predictions, and Richard Harvey was ridiculed when the world did not, in fact, end on April 28, 1583.[122] John Harvey was a practicing physician, and some of Gabriel Harvey's scientific books seem to have been in the possession of all three brothers at some point. Nevertheless, Gabriel Harvey's relative ignorance about most of the scientific topics upon which he comments suggests that he did not fully share whatever specialized knowledge his brothers possessed.

Unlike Richard Harvey, Shakelton had the good sense to keep his predictions of doom vague, arguing that various celestial and meteorological events were caused by God as a call to repentance but avoiding any claims about specific future events. It makes sense that a minister would avoid specific predictions because so-called "judicial astronomy," which involved the prediction of future events based on the configuration of the stars, was considered to be an unacceptable infringement of God's providence. Shakelton argued that the events and discoveries of the 1570s and 1580s provided evidence that the world was growing old, nature itself was becoming unstable, and the world was approaching an end. He had a clear understanding of orthodox Aristotelian naturalistic accounts of nature and also had read and at least partly understood Digges's account of the location of the supernova, repeating Digges's idea that the star may have a supernatural cause.

Shakelton's treatise is particularly interesting because it combines a religious intent with knowledge (and basic understanding) of recent scientific discoveries. He was for the most part an orthodox Aristotelian who did not believe, on religious grounds, in prying into the secrets of nature. His evidence for the decline of the world includes what we would consider respectable scientific thought (his allusions to the Diggeses and other "mathematiques") as well as references to reports of preternatural phenomena like the appearance of armies and lions in the sky. He informs his readers that God has used "this burnyng Comete, whiche now appeareth: as also by the late dreadfull earthquake, but most chiefly by that supernaturall starre of all others that was seen in Cassiopeia . . . to move and stirre up by all the sorte of us (beeyng most wretched and miserable sinners) to be mindfull of the Judgment daie." He says he will not predict the actual date of the end of the world because it is wrong to be "seeking curiously into the secretes of almightie God" (B1v).

Shakelton is aware of the Aristotelian argument that comets are caused by vapors from the earth which rise up close to the regions of fire and are set on fire; as such they were believed to be sublunar phenomena: "I am not ignorant, how Aristotle and divers other Philosophers, dooe define a blasyng Starre, to bee but an yearthly vapour, somewhat thicke, and clammie, and fatte in substaunce, whiche being drawne up, by little and little, from the bowelles of the yearth, into the upper regions of the aire (nigh to the Sphere of, the fierie element) is then set on fire" (C3v) He acknowledges that Aristotelian philosophers therefore believe that "blasyng Starres dooe spring of naturall causes," (C3v) but goes on to make the expected claim that "almightie God, is not onelie the creator, and governour of Nature it selfe, and of all Naturall causes" but also of the star (C5v).

Shakelton goes on, however, to bring Digges's treatise on the star explicitly to bear as proof that it could not be a comet with natural causes but, since it is located in the realm of the fixed stars, must represent a miracle, albeit one with mathematical proof: "For in the judgement of the most experte Astronomers (and suche as are moste skillful in the Mathematiques in this our tyme and age) the

Starre, which did then appeare unto our sighte, was no Comete, but mere supernaturall, directly apposite to the Starre, whiche appeared at the Nativities of our Saviour Christe" (D3v). Evidence for this position "more at large appeareth in the Booke, which was written by that worshipfull gentleman maister Thomas Digges, called Alae seu Scalae Mathematicae, where the Miracle is by demonstrations Mathematicall plainly and most manifestly proved" (D3v).

In its negotiation of orthodox Aristotelianism, new scientific knowledge, superstitious lore, and religious belief, Shakelton's treatise provides a good example of late sixteenth-century attitudes toward the natural world. As evidence for the decline of the world, he adduces "the Mathematiques," who argue that "ye constitution of the celestiall worlde, is not the same that it hath been in tymes past" (C3v), but he also gives a long list of eclipses and comets, and finally states that "in Helvetia in Germanie, were sene in the aire two hostes of men of warre, and two Lions grievously and terriblie fighting together" (C6r). However, he immediately questions reports of this kind: "yet in these daies it will not sincke into the heades of many, that there maie any suche strange sight appeare in the aire, because wee can not render a naturall reason thereof" (C6r).[123] For Shakelton, phenomena either have a "natural" explanation in Aristotelian terms or else they are "supernatural" events caused by God. Even "natural" events, however, can be seen as under the control of God as efficient or final cause and therefore as signs that call for repentance. Shakelton takes the new science, then, not as representing the end of the Aristotelian universe but as presaging the end of the universe itself.

In comparison with Shakelton, Gabriel Harvey seems much more sanguine about the new ideas concerning the nature of the universe of which he was aware, and it seems likely that he failed fully to understand their implications. Spending any time reading Harvey's letters and marginalia makes clear that his contemporaries' dislike was probably justified: he is clearly a pretentious name-dropper. Although some writers like Francis Johnson try to blame Harvey's difficulties at Cambridge on his anti-Aristotelian views, Virginia Stern acknowledges that personal issues had a role in Cambridge University's attempted refusal to confer his master of arts degree.[124] Harvey doesn't hesitate to deliver pronouncements on things he doesn't know much about, and he seems to want to be on the cutting edge of learning, so he makes reference in his marginalia to the works of Digges, Recorde, Harriot, Copernicus, and Dee. Perhaps because of his brothers, Gabriel Harvey seems to have had a greater interest in and willingness to comment on current natural philosophical ideas than other writers from the period. His friend Edmund Spenser seems to have been more reticent about commenting on things he did not fully understand, but some of Harvey's specific ideas about natural philosophy might indicate what Spenser knew, given that they were friends. And, despite a rather catty note in Harvey's marginalia complaining that Spenser didn't know much about astronomy, there are indications that he knew as much as Harvey and probably understood more clearly the implications of what he knew.[125]

Harvey nevertheless provides an important example of how new ideas took form in the popular imagination during this period of ferment and confusion. The forms these ideas assumed frequently do not much resemble what we think of as "the new science," but they register its effect in sixteenth-century terms. Harvey first weighs in on natural phenomena in a long letter to Spenser discussing the earthquake of 1580. On April 6, around 6 p.m., an earthquake, centered in Dover, was felt throughout England and northern France. A number of writers hurried to publish accounts of and responses to the quake, mostly interpreting it as a sign of God's wrath at man's sinful behavior and of the impending end of the world.[126] As Shakelton noted in his treatise, the 1580 earthquake was an unusual event in England and like the recent comets and the supernova was subject to widespread discussion and interpretation. Harvey, on the contrary, seems unperturbed by the earthquake and uses it largely as a pretext for what amounts to a rhetorical exercise. The letter reveals his knowledge of Aristotelian "naturalistic" explanations from the *Meteorology* (to which he refers) and provides an orthodox account of how the earthquake could be both a natural meteorological phenomenon and a warning from God. Harvey's account of the earthquake shows the simultaneous presence of competing "Aristotles" and attitudes toward Aristotle, a critical view of correspondences and resemblances, a fairly sophisticated discussion of the relationship between natural and supernatural causes, and an awareness of epistemological issues. Harvey's letter evinces an epistemology based in "experience" afforded by observation of nature and by authoritative books. It also reveals that criticism of university scholasticism (in Harvey's case from a Ramist perspective) did not entail a rejection of the basic Aristotelian theories of nature. He clearly differentiates a naturalistic discourse based on Aristotle's *Meteorology* from the Aristotle of university disputation, which he ridicules. He also provides a rational and sophisticated consideration of the relationship between natural and supernatural phenomena, and he articulates the proper use of evidence from ancient texts as a way of accessing "general experience."

This letter is part of the published correspondence of Gabriel Harvey and Edmund Spenser, a collection designed to attract attention to the wit and rhetorical skills of these two ambitious friends. Harvey writes as a Cambridge-educated scholar who was in the late 1570s pursuing an academic career. He considered himself to be forward thinking and advocated a reformed university curriculum that would replace the system of disputations based largely on Aristotle with the Ramist method of analysis.[127] Nicholas Popper has traced Harvey's involvement in the 1590s with a group of London-based artisans and instrument makers who were interested in developing "a method of garnering information by deploying technical knowledge."[128] Harvey's way of thinking was not shared by everyone in sixteenth-century England, but it was more common—especially in learned circles outside the universities—than has been acknowledged. He does, however, seem to have been prone to exaggerating his knowledge of mathematics and natural philosophy. In fact, his ignorance of the implications of Copernican theory or the supernova of 1572 meant that he was largely immune to the epistemological angst that afflicted educated contemporaries.

The letter begins with Harvey's description of a card game, involving himself, a gentleman, and two gentlewomen, which is interrupted by the earthquake. Kendrick Prewitt has argued that the social setting is "clearly contrived," probably inspired by Castiglione's *Book of the Courtier*, to showcase Harvey's "oratorical performance, off the cuff, of two explanations for [the earthquake]," which also seem "rather too planned to be plausible."[129] Harvey is not writing a natural philosophical treatise but rather showing off his command of disputational and rhetorical skill. At the request of his host, he actually provides three different explanations for the earthquake: a brief naturalistic account, a more elaborate account presented as a parody of university learning, and a final "short, but sharpe, and learned Judgement of Earthquakes" intended to demonstrate a more acceptable (partly Ramist) presentational method.[130]

It is important to note that, despite Harvey's satire of university Aristotelianism, all three of his accounts are fundamentally based on Aristotle's treatment of earthquakes in the second book of the *Meteorology*. This basic tenet of Aristotelian naturalism is assumed throughout to be true; the gentleman host is already familiar with it when he asks Harvey for further explanation: "May there not be some sensible Naturall cause therof, in the concavities of the Earth it selfe, as some forcible and violent Eruption of Wynde, or the like?"[131] The host is grounded in Aristotelian naturalism, asking for a "sensible" (accessible to common experience) and "Naturall" cause. As Prewitt notes, Harvey's long-winded parodic answer and his "short, sharpe" model answer are both based in Aristotle's *Meteorology*.[132]

Harvey begins with a matter-of-fact statement that recent heavy rains have caused "the poores, and ventes, and crannies of the Earth . . . [to be] so stopped, and fylled up every where with moisture, that the Exhalations, and Vapors, pent up as it were in the bowels thereof, could not otherwise get out, and ascende to their Naturall Originall place" (44). Here, the one reference to a correspondence between the earth and the human body ("bowels") is explicitly identified as a metaphor ("as it were"). When the gentleman asks for an explanation of the earthquake that calls on "your deepe Universitie Cunning" (44), however, Harvey provides a parody of a university disputation.

His mocking account elaborates the basic naturalistic explanation in ways that illustrate his ability to distinguish rhetorical play from serious natural philosophy. First of all, the parodic account is full of jargon associated with the Aristotelian disputational practices of the universities: the Earth must contain "very great store of substantiall matter" comprised of "sundry Accidental humours, and fumes, and spirites, either good, or bad, or mixte" (45) in order "to avoide the absurditie of *Vacuum*" (44). Here, familiar terms from the *Physics*—"substantiall," "vacuum," "accidental"—are combined with a humorously vague list of what the Earth contains: not "exhalations" and "vapors," as understood in the *Meteorology*, but "humours, and fumes, and spirites." When Harvey asserts that his parodic account is based on "Reason, and Philosophy," one of the gentlewomen responds that she "can neither picke out Rime, nor Reason, out of anything I have hearde yet" (46–47). Harvey at another point defends an argument by saying that some "of our finest conceited heads defend this Position, (a very straunge Paradox in my fancie)" (48), suggesting that university disputations are rhetorical flights of fancy rather than serious investigations of truth.

The parodic account is also shot through with the kind of analogic thinking—the correspondence between microcosm and macrocosm, for instance—that scholars, including Michel Foucault, have seen as a constitutive feature of natural philosophy before the scientific revolution. Harvey here clearly differentiates this kind of logic from a naturalistic account of an event like an earthquake, seeing analogy instead as providing the occasion for fanciful rhetorical elaboration. He begins the parody by exaggerating Aristotle's analogy between the earthquake and the human body. Earthquakes occur when conflict between good and evil matter within the Earth is "fostered with aboundance of corrupt and putrified Humours, and ylfavoured grosse infected matter, that it must needs, (as well, or rather as ill, as in mens and womens bodyes) brust out in the end into one perilous disease or other, and sometime, for want of Naturall voiding such feverous and flatuous Spirites, as lurke within, into such a violent chill shivering shaking Ague, as even nowe you see the Earth have" (45). Aristotle's theory of trapped vapors does, of course, involve an understanding that humors could generate "wind," which, when trapped within the human body, could have similar effects: elements and humors were thought to be related substances. What Harvey ridicules, though, is the loose conflation of "humour," fume," "spirit," and "vapor," as if they were identical phenomena, and the construction of an explanation entirely in terms of the analogy. The image of the Earth farting takes the analogy to ridiculous extremes. He underscores the problematic status of this kind of analogical thinking when he goes on to predict (with clumsy innuendo) that the gentlewomen, having heard his explanation, will "this night, within somewhat lesse than two howers and a halfe, after ye be layed, Dreame of terrible straunge Agues, and Agonyes as well in your owne prettie bodyes, as in the mightie great body of the Earth" (47–48).

The logic of resemblance comes in for further ridicule when Harvey claims that some university scholars argue that an earthquake results when the Earth is drunk: "having taken in too much drinke, and as it were over lavish Cups, (as it hath sensibly done in a maner all this Winter past) now staggereth, and reeleth, and tottereth" (48). Other "Contradictorie fellowes" involved in disputation might explain an earthquake as "a certaine vehement, and passionate neesing, or sobbing, or coffing" of the Earth (48). Building to a climax of absurdity, he argues that "the soundest Philosophers in deede and deepest Secretaries of Nature" explain an earthquake as caused by animals fighting underground: "Woormes, and Moules, and Cunnyes, and such other valiantly highminded Creatures, ye Sonnes and daughters of *Mars and Bellona* that nourish civill debate" (49).

Harvey's serious account comes in response to the host's request that he explain whether he considers "Earthquakes, as Naturall, or Supernaturall motions." In Protestant England in this period, there was a lot of discussion about whether meteorological events (comets, earthquakes, strange atmospheric events) were simply natural phenomena as explained by Aristotle or whether they were supernatural events caused by God to foretell dire events. Many English Protestants were invested in believing that God no longer intervened in earthly affairs, so in the last quarter of the sixteenth century, they tended to subject many more events to naturalistic explanations.[133] Harvey's discussion resembles others from the period and reveals clear, rational, and sophisticated thinking about the relationship between the natural and the supernatural. He uses (as others do) the Aristotelian apparatus of causes to differentiate natural material and formal causes from potentially supernatural efficient and final causes, arguing that even if God intends the earthquake as some sort of warning, human beings are not equipped to understand the specific nature of the warning. He concludes that "an Earthquake might as well be supposed a Naturall Motion of the Earth, as a preternatural, or supernaturall ominous worke of God," and, further, that it is "hard, and almost impossible, for any man, either by Philosophie, or Divinitie, evermore to determine flatly the very certaintie either way" (59).

Harvey finishes the earthquake discussion with a suggestion that outlines the epistemological basis of his rational naturalism. He argues that the question of whether earthquakes are natural or supernatural in origin might be resolved by collecting accounts of earthquakes "credible and autenticall, both olde and newe, divine and prophane, Greeke, Lattine, and other Examples (with discretion and judgement, compiled and compared together)." Having made this compilation and comparison, it might "Appeare by general Experience, and the foresayde Historicall Induction of particulars," whether earthquakes "have ominous and significative Effectes" or whether they are purely natural events. This judgment needs to be based on "Experience" of earthquakes, with textual sources seen as the primary source of evidence. These sources are not to be accepted on the basis of their authority but subjected to "discretion and judgement" about whether they are "credible and autenticall" or not. If someone systematically attempted to correlate the occurrence of an earthquake with succeeding dire events of various kinds, for instance, there might be evidence to suggest whether they are "ominous" or not.

Harvey's natural philosophy in this letter is not at all "new." It is based entirely on a relatively untroubled grounding in Aristotelian naturalism. Comments in Harvey's extensive marginalia reveal that, despite his awareness of Copernican theory and of the mathematical and astronomical writings of John Dee, Robert Recorde, and Thomas Digges, he never understood how these new ideas threatened the Aristotelian status quo. Harvey seems to have believed that Copernicus simply reformed Ptolemaic astronomy, making more accurate observations of the location of the heavenly bodies.

Because his marginalia provide no extended engagement with natural philosophy, the extent of Harvey's scientific knowledge is debatable. Francis Johnson takes Harvey's engagement quite seriously, commending his "eager curiosity in all matters pertaining to science" that "caused him to seek out the works and the company of the ablest scientists of his day."[134] Johnson exaggerates the extent to which Harvey's Ramism was truly anti-Aristotelian and ties his interest in new ideas to Ramist thought. Stern emphasizes the role of Harvey's brothers, Richard the astrologer and John the physician, in involving him with specialized knowledge of the natural world.[135] Nicholas Popper argues that Harvey became interested in science after he moved to London, becoming part of "a group of scholars in London in the 1590s such as Francis Bacon and Hugh Plat, who were examining the benefits local artisans might offer the commonwealth and reconsidering the intellectual status of craft knowledge."[136] Popper argues that Harvey's move from London to Cambridge in 1590 and his interest in Ramist reform made him more aware of the necessity of "useful knowledge" and the importance of those who used "instruments and craft knowledge to garner reliable and useful information."[137]

Popper acknowledges that Harvey's engagement with craft knowledge and natural philosophy in his marginalia usually takes the form of lists: lists of authors, artisans, and books pertaining to a given topic. Popper sees Harvey's penchant for providing lists of recommended authors as "an exercise in humanist copia" designed to "bring prestige to the categorizer."[138] While these lists may be largely exercises in name-dropping, they do provide evidence of Harvey's partial knowledge of current developments in natural philosophy. Like many of his contemporaries, Harvey often seems to be struggling to understand the implications of his new knowledge. For instance, he sometimes lumps together thinkers with quite different philosophical positions. A list of those who have questioned Ptolemaic cosmology includes "Regiomontanus, Cardanus, Copernicus, Reinhold, Apian, Joachim Rheticus, Gemma Frisius, Jofranc Offusius, Statius, Maestlinus, Tycho Brahe, and many others"; it is a list that includes advocates of both the geocentric and heliocentric models.[139]

However, Harvey's marginalia reveal that someone whose life intersected with writers like Sidney and Spenser, Nashe and Greene, was aware of a wide range of writing about the natural world, although that knowledge was eclectic and incomplete. Although he refers to recent ideas, it is usually in marginal notes to books that are quite old-fashioned. As was the case in the earthquake letter, Harvey often seems concerned with natural knowledge as an occasion for rhetorical or poetic elaboration rather than for its own sake.

In the marginal comments Harvey wrote in Blagrave's *Margarita*, he notes that he knows Blagrave and Digges personally and they have recommended a specific instrument maker to him (211). Others have recommended other artisans to him who make "feet, frames, rulers, compasses, and squires," as well as astronomical instruments (211). This book also contains Harvey's comparisons between London artisans and university scholars: "An Youth: & no university-man. The more shame for sum Doctors of Universities, that may learn of him" (212); "Schollars have the bookes: & the practitioner the learning" (213).

Harvey makes comments in his marginalia that some have taken as espousing a nascent "pre-Baconian" empiricism, but he seems still pretty firmly in an Aristotelian mode: "All sciences are based on the senses and reason, clearly they accord with reason, and with sense. Reason is the soul of each of the principles. Experience is the soul of souls, the firmest demonstration and irrefutable criterion. Give me visible and rooted demonstration of each principle, experiment, geometric instrument, astronomical, cosmographic, horologiographic, geographic, hydrographic, or mathematical in any way."[140] The important word in Harvey's statement is "Experientia," since Aristotelian "experience," as we have seen, involved knowledge of the manifest, everyday operations of nature. Harvey goes on to include "experimenta" as one form of experience, alongside data from instruments of various kinds. But it is not at all clear that he views "experimenta" as producing anything other than data about ordinary experience. His commentary then, typically, trails off into a list of impressive sounding areas of study.

Harvey explicitly refers to Copernicus a couple of times, although he never indicates what he understands about the geocentric model of the universe. As noted above, he includes Copernicus in a list of astronomers who have questioned the Ptolemaic system, but that list includes mostly writers who accept the heliocentric model. In a marginal note in his copy of *Firmini Repertorium De Mutatione Aeris* (a 1539 edition of a prognostication written in the late fourteenth century), Harvey links the phenomenon of precession of the equinoxes to Copernican theory: "Changed judgements: since the fixed stars have changed places. The old tables of Ptolemy and others are not congruent with today's observed phenomena: it is very evident from the Copernican revolutions and the Prutenic tables of Reinhold, where the errors of the old tables ought to be reformed."[141]

It isn't clear from this comment whether Harvey understands the issues at stake. Precession, as noted above, describes the fact that the constellations, supposed to be fixed and immovable, were observed in antiquity to have shifted their position in the sky about one degree every hundred years. However, these stars only appear to have moved from a Ptolemaic perspective; from a Copernican perspective, precession is explained by a wobble in the earth as it rotates. Harvey seems to imply that Copernicus has simply corrected the positions of the constellations, reflecting their movement since the time of Ptolemy (which is what Firminius advocates), whereas, in fact, Copernicus argued that the stars were not moving at all. Spenser, in Book V of *The Faerie Queene* and also in the *Mutabilitie Cantos* seems to link precession with movement of the fixed stars that implies that the universe has decayed, the conclusion usually drawn from Digges's demonstration that the supernova of 1572 was located in the sphere of the fixed stars. It seems possible that Harvey and Spenser share the same confusion about precession and its relation to new ideas about the natural world. Harvey does not, however, share Spenser's concern about the idea that the fixed stars can change.

Harvey's marginal comments about medicine are similarly eclectic. His ideas seem to be for the most part in line with orthodox Galenic humoral medicine: "If the disease be hott, & drie, use remedies cold, & moist."[142] In this volume he criticizes "these new writers, especially the Germans" as being "more compendious, than subtle" (127) and seems to side with Aristotle and Galen against the Paracelsans when he notes that "Nature is the best interpreter of herself. She lays open a physiognomic school in all of her works," suggesting that the signs provided by nature are obvious and open, not secret (128). However, in his comments on the Paracelsan writer Braunschweig, he notes that Braunschweig's own long life suggests the usefulness of his medical ideas, and he now notes approvingly that the book offers "secreta therapeutica" (131). However, he later mentions that the book was recommended to him as an introduction to Greek medical writers like Galen (132). In another book he notes that "Paracelsus considers Galen to be a sophist; Ramus thinks the same thing about Aristotle" (155). Harvey doesn't take sides here, but we have already seen that he could espouse Ramism while still maintaining many Aristotelian principles, and the same seems to be true with regard to medicine. He may not fully understand how Paracelsan medicine differs from Galenic theory; like many people at the time, he embraced some of the Paracelsan chemical remedies without fully buying the whole theoretical paradigm.

Clearly, some educated Londoners like Shakelton and Harvey were aware of the writings of Digges, Dee, and Recorde, and therefore of Copernican theory and the location of the supernova. The default view of the natural world in late sixteenth-century England was a naturalistic Aristotelianism grounded in the *Meteorology*, popular versions of Ptolemaic astronomy, and Galenic medicine. Shakelton and Harvey believed in the possibility of supernatural events but carefully delineated a relationship between natural and supernatural. Both were aware that the Aristotelian/Ptolemaic/Galenic model had been shaken in recent years. Shakelton was more typical in his response; Harvey seemed to share the optimism of Digges without necessarily understanding its source. Like other writers who grappled with these issues in late sixteenth-century England, Shakelton and Harvey were concerned with epistemological questions: Is the universe intelligible? Does it reveal itself openly, or through secret signs? Are the data provided by the senses accurate? If they are not, what recourse is there? In the next chapter, I look at the surprisingly sophisticated discussions of these issues found in texts throughout the sixteenth century.

CHAPTER 3

Losing Touch with Nature

In chapter 2, I cited treatises published in the vernacular and in Latin, written for the most part by experts with specialized knowledge of mathematics, astronomy, and/or medicine and often designed to convey up-to-date knowledge about nature to a wider public. All these books demonstrate a concern with the relationship between sensible daily experience—the intuitive view of nature codified in popular Aristotelianism—and the increasingly counterintuitive theories emerging from the continent or from experience of nature itself. These treatises reveal four distinct stages in the evolution of attitudes toward the intelligibility of nature. Early works (written before the 1540s) are for the most part confident that ordinary experience of nature—aided, perhaps, by some instruction and the use of specialized instruments—enables access to truth. Nature reveals its causes on the surface. As the century progressed, writers became less confident about the reliability of the senses and more aware that ordinary experience of the natural world does not always provide access to truth about it. In the 1540s and 1550s, writers acknowledged that human beings often cannot have direct access to nature's truths and as a result need instruction in how to read the "signs" and "tokens" that appear on the surface.

For the members of the Dee circle who understood the implications of Copernicus's *De revolutionibus*, the discrepancy between the appearance of the universe from Earth and his theory about its true configuration eroded their confidence that they could teach people how to reach the truth about nature by interpreting its signs. In the 1550s and 1560s, these writers pursue the possibility that the study of optics might explain why the universe presented such a deceptive appearance (as a kind of optical illusion) or else might lead to the development of devices that would allow them to see past its illusory surface. However, the supernova of 1572 seems to have ended their hope that the unsettling of the Aristotelian system was just an illusion. As a result, there was increasing interest in the various "secrets" traditions that purported to offer access to hidden ("occult") knowledge through various adept practices; these traditions included astrology, alchemy, Paracelsanism, Neoplatonism, natural magic, and advanced mathematics. By the end of the century, many writers (such as John Dee, Thomas Digges, Francis Shakelton, Gabriel Harvey, and William Gilbert) called on both the naturalistic and secret traditions.

Today we may find these writers' approach to epistemological questions surprisingly sophisticated. Sixteenth-century discussions of natural philosophy do sometimes question the authority of antiquity. Even quite early in the century, writers like John Rastell and Thomas Elyot grappled with questions about the reliability of the senses and the ways in which experience can, or cannot, reveal the truth about nature. Some, like Robert Recorde, explicitly warned against accepting the truth of ancient writing about the natural world because of an author's authority. Many writers repeated the commonplace that nature offers "signs and tokens" for us to interpret: however, all of these writers acknowledged that there is uncertainty about what nature's signs reveal and how they are to be interpreted. Writers like Leonard Digges, Dee, Shakelton, and William Fulke, who considered the relationship between natural and supernatural interpretations of natural signs, all offered a fairly complex sense of the relationships between natural and supernatural causes of natural phenomena. These writers recognized, quite early on, the difference between "experience" and "experiment." All of these writers were aware of natural phenomena that can't be readily explained in naturalistic terms. They tended to invoke one of two tropes to convey their means of access to hidden knowledge. In the naturalistic tradition, writers often imagined a figure (such as Nature or Experience) who instructs the student about things that cannot be easily observed. Writers more allied with the secrets tradition were likely to speak of a knowledge or technique that would allow the adept to travel to realms (such as the heavens, or the center of the Earth, or the inner recesses of the body) hidden from most observers.[1]

John Rastell's *The Four Elements* evinces a concern with epistemological issues that would become a preoccupation throughout the century. (It was written very early, probably in 1518 or 1519.) This didactic interlude is designed to inform its audience of basic facts about the natural world, including information about recent voyages of exploration and discovery. It is based on contemporary continental sources, including Joannes Sacrobosco's *Textus de Sphaera* (1511), Gregorius Reisch's *Margarita Philosophica* (1508), and Martin Waldseemueller's

Cosmographiae Introductio (1507), and it has the express purpose of making scientific knowledge from these Latin treatises available in English. Because there are "dyvers prengnaunt wyttes . . . whiche nothynge but englyshe can understand," it is important to translate "connynge laten bokys" so that "all subtell sciens in englyshe might be lernyd."[2] The play provides a basic account of Aristotelian naturalism, including the four elements, the relationship between the sublunar and heavenly spheres, and the position of the Earth in the Ptolemaic cosmos, as well as a survey of cosmography that includes references to the New World. It presents all of this information as based not in the authoritative texts from which Rastell gleaned it but instead on observation of the natural world provided by "experience," "instruments," and "experiments." There is a real concern with the relationship between knowledge that is manifestly evident (like the behavior of the elements on Earth) and knowledge that seems to contradict the evidence of the senses (like the fact that the Earth is spherical). Here, as in the later works by Robert Recorde, the use of dialogue allows the student's doubts and skepticism to be voiced alongside the instruction provided by various teachers.

Rastell links his justification for writing the interlude to the importance of disseminating knowledge about the visible world: "How dare men presume to be callyd clerkys, / Dysputynge of hye creaturis celestyall, / As thynges invysyble and Goddes hye warkys, / And know not these vysyble thyngys inferyall?" (ll. 113–16). Rastell's play takes the form of a series of dialogues between Humanyte and various figures who instruct him (or tempt him away from instruction, following the usual plotline of a Tudor interlude). Although the Messenger who introduces the play emphasizes the manifest nature of the four elements "whose effectis dayly appere here at eye" (l. 125), Humanyte repeatedly raises questions and expresses doubts about facts that are not easily visible and that seem unreasonable because they contradict the evidence of the senses. Humanyte "can no maner wyse parceyve nor see, / Nor prove by reason why the yerth sholde be / In the myddys of the fyrmament hengyng so small / And the yerth with the water be rounde with all" (ll. 336–39). Later, he insists that "yet it provyth not right / That the yerth by reason shulde be rounde" (ll. 363–64). Studyous Desire promises to bring in Experyens to prove to Humanyte that the earth is round: "Experyens, / Of dyvers instrumentys is never without, / Cowde prove all these poyntys, and yet by his scyens / Can tell how many myle the erthe is aboute" (ll. 391–94). Experyens's "instrumentys" probably include some form of astrolabe or celestial sphere, since he appears onstage carrying a "fygure" (l. 690), which is probably a globe of this type. However, his knowledge is based in orthodox Aristotelian "playne experience" (l. 1041). He judges the correctness of a map because "bothe the se and lande also / Lye trew and just as the sholde do, / I know by experyens" (ll. 693–95). Experience does also resort to "myne instrumentis" and "playne experimentis" (ll. 1122, 1123) to convince a skeptical Humanyte that the world is round.

Despite the reliance on Experyens as a source of information about the natural world, there are also dangers associated with it in Rastell's play. It is probably not a coincidence that when Studyous Desire first promises to bring in Experyens, Sensuall Desire appears instead and lures Humanyte away from the serious study of nature for several scenes. Because Experyens is based on evidence of the senses, it is problematically close to sensuality. The play as we have it is a fragment, and the section where Experience does "experimentis" with his "instrumentis" is missing. Instead, after the gap, we find Humanyte once again consorting with Sensuall Appetyte and Yngnoraunce. Humanyte goes into hiding, "for I was almoste for fere / Evyn clene out of my mynde" (ll. 1201–2) as a result of the violent antics of his two companions. However, Sensuall Appetyte insists that "it is the study that ye have had / In this folyshe losophy hath made you mad" (ll. 1203–4). Humanyte has either been driven away from learning or frightened half to death by knowledge gained through instruments and experiments, which counters the evidence of the senses.

Thomas Elyot's *Castel of Health*, probably published around 1533, provides another early example of a text written to provide instruction in basic science to a lay audience, in this case offering an introduction to Galenic medicine and a "regimen" designed to preserve health.[3] Like Rastell, Elyot is basically confident that our daily experience of the body and its symptoms gives access to reliable knowledge about the causes of disease as manifested by changes in temperature and bodily fluids. However, as in Rastell's play, Elyot sometimes acknowledges that what we observe needs to be explained or interpreted, in this case because the interior of the living body is hidden from view. Looking ahead to the language of later treatises, Elyot talks about "signs" and "tokens" that need to be read by a trained practitioner.

In defending his book against physicians who might object to making medical knowledge available to lay readers, Elyot argues that he is providing people with information so that "the uncertayne tokens of urines, and other excrementes, shuld not decyve them, but that by the true information of the sycke man, by me instructed, they mought be the more sure to prepare medicines convenient for the dyseases."[4] Here Elyot's point is that a layman armed with basic medical knowledge would possess information about the conditions that might alter the appearance of a specimen he provides to his doctor: the time of day that the specimen was produced, for example. However, this justification entails the admission that the humoral fluids produced by the body provide "uncertayn tokens" that can be deceptive to untrained observers.

Elyot's sense that humoral "excrements" require interpretation may stem from his reading of Aristotle's theory of the elements. Elyot offers a basic account of the Aristotelian elements and their relation to the bodily humors, and he is aware of the medieval controversies over the nature of mixtures and therefore suggests that human experience of the elements is at one remove from their essential nature: "none of the sayd elementes be commonly sene or felt of mortall men, as they are in theyr orygynall beynge: but they whyche by our sences be percyved, be corrupted with mutuall myxture, and be rather erthy, watry, ayry, and fyry, than absolutely erth, water, ayre, and fyre."[5] Differing slightly from Rastell's sense that the elements are what we "dayly see here at eye," Elyot acknowledges that we don't have access to the elements in their pure form but only as mixtures.

Nevertheless, Elyot seems confident that states of health and disease can be diagnosed through visible "signs." He defines "complexyon" as "a combynation of two dyverse qualities of the iiii elementes in one body," so that, for example, "the body where heat and moisture have soverayntie" is called "Sanguyne." A sanguine complexion "is perceyved and knowen by these sygnes" including "carnositie of fleshynes," "the veynes and arteryes large." Similarly, disease is caused by "distemperaunce," which "bryngeth sycknesse or griefe into the member. Wherefore to knowe the distemperature, these sygnes folowyng wold be considered" (3r). As in Rastell, nature makes its causes manifest, although special knowledge is sometimes needed to understand what is being observed.

From the 1640s through mid-1650s, popularizing treatises on mathematics, medicine, astronomy, and meteorology continued to promulgate the basic tenets of Aristotelian naturalism and Galenic medicine.[6] Writers connected with the circle of John Dee emerged in midcentury as the group in England most aware of current developments in natural philosophy and, perhaps as a result, most attuned to epistemological questions. (I look at works by Robert Recorde on mathematics, medicine, and astronomy, and a treatise on meteorology and astrology by Leonard Digges.) All of these texts reveal their authors' awareness that nature does not always reveal its truths clearly and unproblematically; some form of interpretation is required. Regardless of focus, these treatises seek to instruct readers in the proper ways to read the "signs and tokens" found in nature.

These references to signs and tokens are related to the tradition of the "book of nature" traced by James Bono, Peter Harrison, and others but differ from it in that they refer to signs that nature provides about the causes of its own workings rather than providing moral instruction for human observers.[7] These midcentury treatises are not explicitly describing nature as book, or as offering up a Ficinian correspondence between language and what it names, or Paracelsan "signatures" imprinted in nature by God.[8] The "signs" mentioned in midcentury medical, mathematical, and astronomical texts appear to have had two primary sources, one Aristotelian and the other biblical. Craig Martin has argued that Aristotelian meteorology was not open to "demonstrative proof" because it dealt with the imperfect and unstable mixtures of the elements.[9] As a result, knowledge about meteorological phenomena is based on "application of signs that confirm or suggest a possible theory."[10] Aristotle uses the words *semeion* and *tekmerion* to refer to observed meteorological events that provide evidence about their nature and cause.[11]

Signs and tokens are also found in the Bible, for example, in Genesis 1:14: "And God said, Let there be lights in the firmament of the heaven to divide the day from the night; and let them be for signs, and for seasons, and for days, and years." This verse is cited in discussions about the validity of judicial astrology but seems to have been more broadly interpreted to suggest that nature offered up signs of its own workings. In translating this passage, the Tyndale Bible reads "signs" and the Coverdale Bible reads "tokens," so the phrase "signs and tokens" may also refer to these two early versions of the passage. In midcentury treatises about the natural world, "signs and tokens" can be found not only in the heavens but in mathematical equations, in urine, and in earthquakes and clouds. Signs and tokens seem to be a way to name any natural phenomenon that requires interpretation to discern its causes or true nature. Signs are manifest and perceptible in the course of daily experience yet require some kind of training in order to be properly interpreted. They mark a kind of halfway point on the continuum between the manifest and the occult or hidden.

The question of how to read the signs and tokens of celestial or meteorological phenomena led inevitably to a discussion of whether they give access to knowledge that is natural or supernatural. At this point in the century, God was widely recognized as the "final" cause of all events, so that any natural occurrence could be a sign of divine approbation or disapprobation. Later in the century, Protestant concern to show that the age of God's miraculous intervention in nature had ended led to a writer like William Fulke who found natural causes for virtually all strange atmospheric and celestial events. Even at midcentury, though, Leonard Digges recognized that signs—especially in the air or heavens, could have supernatural and natural significance at once. Digges recorded "marvelous" occurrences such as armies appearing in the sky, but he clearly differentiated between phenomena that he actually observed and questionable phenomena that he had simply read about. Recorde insisted even more explicitly than Digges that knowledge of natural phenomena should be based on experience and not on the authority of ancient writers.

1530 TO 1560: SIGNS AND TOKENS

Robert Recorde was a Welsh physician and mathematician who was educated at Oxford and became a fellow of All Souls in 1531. He received a medical degree from Cambridge in 1545 and was general surveyor of mines and money from 1551 to 1558.[12] Recorde is best known for a series of popular textbooks: *The Ground of Artes* (1542), a treatise on arithmetic; *The Pathway to Knowledge* (1551), on basic Euclidean geometry; *The Castle of Knowledge* (1556), on astronomy and the armillary sphere; *The Whetstone of Witte* (1557), an introduction to very basic algebra; and a medical treatise, *The Urinal of Physick* (1547). Recorde's emphasis on the importance of basing knowledge on experience and observation rather than authority has led some scholars to identify him as "anti-Aristotelian," but Recorde seems pretty firmly entrenched within the tradition of Aristotelian naturalism.[13]

Recorde's *The Ground of Artes* reveals that even basic arithmetic in this period underwent a change that contributed to the increasing abstraction of the scientific understanding of the world from an older concrete essentialism.[14] As a result, Recorde's treatise on arithmetic alludes to some of the same epistemological issues that can be found in his works on astronomy and medicine. Recorde's volume was the most popular English arithmetic text in the sixteenth century and went through fifteen editions before 1600, with later editions edited and enlarged by John Dee.[15] As noted in the previous chapter, it is largely concerned with introducing Arabic numerals and positional notation as a replacement for Roman numerals, and calculation with pen and paper as a replacement for calculation with counters.[16] It tries to help students move away from theories of ancient mathematics in which arithmetic was tied closely to the enumeration of objects.[17]

Like most of Recorde's books, *The Ground of Artes* is in the form of a dialogue, so the student's questions about the nature of knowledge and representation are directly articulated. Calculation with Arabic numerals forces a separation between the numeral and the value that it represents; thus the concept of "betokening" or signification enters the realm of mathematics. As Alfred Crosby has suggested, use of a counting board "bestowed the advantages of both place value and zero without the disadvantage of having to think about them."[18] When the Master sets out to teach "the Art and use of Arithmeticke with Penne" and introduces Arabic numerals, the Scholar expresses confusion about place value, querying the Master about the "difference between the value and the figures."[19] The Master explains that unlike Roman numerals, where a figure like X always represents 10, Arabic numerals work differently: "for the value is one thing, and the figures are another thing: and that commeth partly by the diversitie of figures, but chiefly of the places wherein they be set" (43). Therefore, he explains that "everie figure hath two values: one alwayes certaine that it signifieth properly which it hath of his forme: and the other uncertaine, which he taketh of his place" (43).

Given the example of the numeral 3,679, the Scholar is still confused about the relationship between the appearance of an individual numeral and the value that it signifies: "doeth not 7 that standeth in the second place, betoken vii? And 6 in the third place betoken vi?" (46). The Master needs to explain again that "in the first place, everie figure betokeneth his owne value certaine onely, so in the second place everie figure betokeneth his owne value certaine ten times" (46). Value is still "certaine," but the same numeral can betoken different values in different places. The Scholar then expresses confusion about the new figure of zero, and the Master explains that "they are of no value themselves, but they serve to make up number of places" (49).

Recorde's text was especially useful to merchants who needed to learn basic arithmetic and algebra in order to carry out business transactions.[20] These readers would thus have been introduced to a rupture at the most basic level of correspondence between number and thing: numerals no longer had a fixed value but could "betoken" different values in different places. Even worse, a new figure, zero, is introduced that "doth signifie nothing" (43) and is "of no value" in itself (49). Mathematics eventually became, for Galileo, the language in which the book of nature was written; however, the movement toward the new science began with what seemed to be a widening gap between the concept of number and the things enumerated, between math and the physical world.

Recorde's medical treatise *The Urinal of Physick* (1547) was, like Elyot's *Castel of Health*, intended for a lay audience.[21] Its topic is more specialized, focusing only on the interpretation of urine. Although he is firmly in accordance with humoral physiology, Recorde seems more aware than Elyot of the gap between the processes going on inside a living human body and the symptoms manifested in urine. He takes pains to explain the relationship between what is happening inside the body and the "tokens" visible in urine. The whole book is essentially about the ways in which urine can "betoken" or "sygnify" various states of health or sickness.

Near the beginning, Recorde makes clear the difficulty of getting readers who know nothing of anatomy to understand "the generacion of urine without sum sensible example" (10r). He proceeds to give two analogies that provide a "sensible example" of this hidden process, expressing a sense of the limited role of analogy in science that Dedre Gentner and Lorraine Daston have associated with later figures like Galileo and Robert Boyle.[22] Recorde carefully explains that "bycause it is very harde to fynde an artificiall example, whiche can a lone duly expresse this worke of nature, I wyll use therefore an example of a natural worke, which shal expresse in menny poyntes this thyng, though not in all: for soche can there none be, but the thyng it selfe" (10r). Recorde proceeds to give two different analogies, each explaining part of the process of the "generacion of urine." First, he uses an analogy with distillation to explain how heat can "separate the purest parte of the juce from the herbes" (10r), just as the kidneys separate waste from the blood. For later writers, distillation will become a means of revealing secret knowledge, but for Recorde it is a manifest practice that can be "daily sene" (10r). He then explains that "as the arte of man useth [distillation] to make this water, so doth nature use to make the water of springes, whereof cum all ryvers, stremes, and flouddes, excepte the see" (10v). The kidneys turn concocted blood into urine, "as the coldnes of the ground turneth ayer into water. But you must take this comparison or similitude too be spoken of the alteration it self and not of the cause," (11v–12r), again emphasizing the targeted nature of the analogy.

It is worth pausing for a moment over Recorde's sense that the production of urine in the body is in some ways like the production of water in the ground, since it reveals that the correspondence between microcosm and macrocosm in the sixteenth century did not necessarily take the form posited by Foucault. For Recorde, analogy does *not* make "possible the marvelous confrontation of resemblance across space," nor does he believe that analogy has "a universal field of application" such that "through it, all the figures in the whole universe can be drawn together."[23] Recorde is very clear that the comparison he draws between the body and the earth is an imperfect analogy, not a mystical correspondence. The Paracelsan doctrine of "signatures" which did link microcosm and macrocosm through mystical connections was not influential in England until later in the century. Something like Foucault's system of correspondences gained force in England only as the Aristotelian transparency of the universe broke down, perhaps as a measure to create a new intelligibility to replace the more straightforward one that was being lost.

Recorde then explains how water in the ground picks up sensible qualities from the hidden materials deep within the earth with which it has contact, bringing those qualities to the surface where they can be perceived: water reflects "by taste and colour, the qualityes of the erth, or vaynes of metal, whereby it runneth and from whence it cummeth" (12r). Urine similarly brings traces of the hidden interior of the body to the surface so that they can be perceived and diagnosed: "so the urine by coloure & other ways, declareth of what sorte the places that it cummeth throw, and humors that it cummeth from" (12r). Thus, if "the blode be pure and clene" and if there is no "grefe in the renes, water pipes, bladder, nor shafte: then shal the urine so declare it, beyng also parfecte and pure in substance and coloure and al other tokens according to the same" (13v). If there is "greefe" or the blood is corrupt, "then shal the urine declare certen tokens of the same" (13v).

For Recorde, the Aristotelian qualities—hot, cold, moist, and dry—if "duly tempered . . . be the cause of contynual helth," and "if they be altered wrongly, then doo they cause dyseases dyversely" (23v). These qualities "cause all alteracions in uryne," so knowledge of them allows the educated observer to "perceyve not onely what every uryne doth betoken" but also "the cause why every urine doth so sygnify" (23v). The balance of qualities within the body are not directly manifested to human experience, but with training, we can learn to read their "tokens" in urine.

Recorde's astronomy text, *The Castle of Knowledge* (1556), contains the first reference to Copernican theory in English.[24] It extends to astronomy and astrology the concern with the relationship between the manifest experience of the world provided by the senses and the actual nature of that world, given particular urgency by the mathematically satisfying but counterintuitive model of the universe proposed by Copernicus. I want to look first, however, at an astronomy text written by another member of the Dee circle, Leonard Digges, and published a year before the *Castle of Knowledge* in 1555 (a revised edition of an early version published in 1553).[25] Leonard Digges was not aware of (or chose not to engage with) Copernican theory. Indeed, his son Thomas Digges published an addendum to his father's text in 1576 replacing his father's "description or Modill of the world . . . according to the doctrine of Ptolomie" with the "new Theorick or Modill of the world" proposed by Copernicus. Despite its lack of engagement with Copernican theory, Digges's A *prognostication everlasting* provides a revealing example of midcentury thinking about how to read the signs provided by the natural world, and it shares with Recorde's work a concern with working out the relationship between astronomy and astrology, between natural and supernatural signs.

As its title indicates, Digges's work is a "prognostication," in Francis Johnson's words "by far the best of the 'perpetual almanacks' issued during the last half of the sixteenth century . . . filled with all sorts of useful astronomical and astrological tables and rules."[26] It includes "astrological and meteorological rules for predicting the weather (the former tabulated at great length); descriptions of various meteorological phenomena and accounts of their causes; tables for calculating the calendar and dates of the movable feasts for several years to come; the rules for bloodletting as prescribed in astrological medicine; and, finally, the description of an instrument for telling the hour at any time of day or night, together with the astronomical tables necessary for its use."[27]

Digges's subtitle reveals the extent to which the volume straddles the naturalistic world of astronomy and meteorology and the more speculative realm of judicial astrology: the *Prognostication* contains "plaine, brief, pleasaunt, chosen rules to judge the weather by the Sunne, Moone, Starres, Cometes, Rainbow, Thunder, Cloudes, with other extraordinary tokens, not omitting the Aspectes of Planetes, with a briefe judgement for ever, of Plenty, Lacke, Sicknes, Dearth, Warres, etc., opening also many natural causes worthy to be knowen."[28] The book is entirely devoted to instruction in how to read signs or "tokens"—and the relationship between signs of weather and signs of impending doom is carefully parsed. Digges is particularly aware of the difference between signs that he has personally observed and signs that are vouched for by the authority of books. The latter are included, but Digges's careful attribution of sources tends to differentiate them from the phenomena that he has himself observed.

The book begins with straightforward tips for predicting the weather: "The sunne in the Horizon or rising, cleare and bright, sheweth a pleasant day: but thinly overcast with a cloude, betokeneth foule weather" (5r). I doubt that sixteenth-century Britons needed a book to tell them that bright sunshine "betokens" a nice day. Digges proceeds to some more complicated prognostications that involve astrological signs: "consider the nature of the signe where the Moone is at the chaunge, quarter, and full. If she be in hote and drie signes, as Aries, Leo, Sagittarius, in winter a good token of faire weather" (5v). Here, the effects of the signs of the zodiac on elemental qualities (hot and dry) are seen as tokens of particular kinds of weather, representing a version of astrology that stays largely within the terms of Aristotelian naturalism.

Digges also extends prognostication to include the prediction of human events by natural signs, but he almost always attributes these prognostications to learned sources and sometimes questions their legitimacy. The naturalistic and the supernatural blend together, as when we learn that "comets signifie corruption of the ayre. They are signes of Earthquakes, of wars, changing of kingdomes, great dearth of Corne, yea a common death of man and beast" (6r). Aristotelian meteorology, of course, held that comets were the result of vapors or "corruption of the ayre" rising up to the region of fire and igniting; these vapors also caused earthquakes and could be associated with bad weather for crops, so about half of these events arise naturally from the conditions that produce comets. As for their prediction of wars and regime change, Digges cites Pontanus, "belli motus, feraque arma minantur, " including further marginal citations of Cardanus and Mizaldus on "de cometarum prodigiis" (6r). He doesn't need citations to support the naturalistic effects of comets but piles them on to support their ability to foretell political disaster.

Digges sometimes questions claims that meteorological events can predict changes in human affairs. Writing about the significance of thunder, Digges notes that "some write (their ground I see not) that Sundayes thunder, should bring the death of learned men, Judges, and others" (6v). Although he defends "judiciary astrology" (the use of astrology to predict human events), he considers it "the least part among a number of his most certayne doings" (7r), and his examples remain in the realm of naturalistic phenomena. When he writes about rain, he provides the traditional naturalistic account: "a colde vapour, an earthly humour, or fumosities, out of waters or earth drawne up by the virtue of the Sunne, to the neather part of the middle space of the ayre, there through cold thicked, then dissolved" (13r). He adds, "Here I leave to speak of miraculous raines, as Milke, Blood, Flesh, Iron, Wooll, etc. For more satisfying in these reade Plynius in the second booke. 58 chapter" (13r). Digges seems to feel the need to include traditional claims about miraculous or prognosticative phenomena, but he distinguishes them from phenomena that lie within his personal experience.

Although the scope of a meteorological/astrological guide like Digges's *Prognostication* seems quite different from that of a textbook like Recorde's *Castle of Knowledge*, the books share a similar attitude toward the status of experiential evidence and the relationship between astrology and astronomy. Recorde's book provides a good example of a midcentury moment when Copernican theory had just become known in England and its implications were still being assimilated. Recorde's book is in dialogue form, in which a Master instructs a Scholar about the basics of Ptolemaic astronomy, using a celestial sphere—a 3-D model of the universe—to explain his points. The book provides a basic description of the elementary and celestial spheres, then uses the "Materiall sphere" to illustrate the movements of the heavenly bodies. The model of the celestial sphere on which the book is based necessarily offers a geocentric perspective, so for the most part the Master instructs the Scholar in orthodox Ptolemaic theory.[29] The Scholar raises questions about the phenomena that are troubling that system—precession of the equinoxes and the retrograde motion of the planets—and is repeatedly told by the Master that these matters are too advanced for him to understand. The

Copernican theory is mentioned in this context, but, as we saw in the previous chapter, the Scholar is again told that he is not adequately learned to assess it. The potential disruption of the geocentric universe, and therefore of the evidence of the senses which had supported it, thus persistently haunt the book, even as it refuses to confront them. In this uncertain moment, the fixity of the fixed stars, the evidence of the senses, and the status of ancient authority become especially important.

Recorde's preface begins by stressing the permanence of the fixed stars: they "waxe not werye with laboure, nother growe olde by continuance, but are as freshe nowe in beutye and shape, as the firste daye of their creation, and as apt nowe to performe their course, as they were the first hower that they began" (A4r). These stars "utterly stande cleere from all corruption of time" (A4v). The unchanging nature of the supralunar sphere had always been an important guarantor of stability in the Aristotelian system, but it may have taken on more importance for Recorde in the wake of Copernicus, since the Copernican model of the universe shored up their fixity by explaining away the slow movement of precession as an effect of the movement of the earth. Recorde briefly alludes to the role of the stars in predicting human events, but he takes the safe position on judicial astrology that the stars do not predict specific happenings but rather generally urge human beings to repentance: "so was there never anye greate chaunge in the worlde, nother translations of Imperies, nother scares anye falle of famous princes, ne dearthe and penurye, no death and mortalitie, but GOD by the signes of heaven did premonishe men therof, to repent and beware betyme, if they had any grace" (A5r). Recorde then dismisses signs of this kind as irrelevant to his book because they "appertain to the Judicial part of Astronomy," which he is not going to treat (A5r).

Although he will not discuss judicial astrology in his treatise, his preface nevertheless calls on the concept of secret knowledge. There is a passage in the preface that appears to be in verse, although it is not typographically delineated as such. This passage addresses God:

> Oh mervailous maker, Oh God of good governaunce:
> They woorkes are all wonderous, thy cunning unknowen:
> Yet seedes of all knowledge in that booke are sowen.
> The signes of the tymes who can them comprise?
> The tokens of troubles what man could devise?
> And yet in that boke they rightly can reade,
> To all secret knowledge it will him straighte leade. (A4v)

In this passage, it is unclear whether the "signes" and "tokens" that can be read in the book of nature are indications of human affairs (wars, fall of princes) or of the kinds of natural phenomena addressed in the body of the book. Recorde seems to indicate that he is referring to "secrets" about the structure of the universe itself. He alludes to Genesis 1:14: "yet dyd God at the beginning ordaine the starres to be as signes and tokens of times alteration: and namely of suche straunge effects as seldome come in ure, and therefore are knowen but to fewe men" (A4r). These "straunge effects" could be wars and regime change but are more likely astronomical phenomena, such as the irregular orbits of the planets or precession of the equinoxes, which were eroding the credibility of the Ptolemaic system (and which could be better explained by the Copernican system). The reason to gain knowledge of astronomy in 1556 would have been to understand a universe in flux.

Indeed, when the dialogue begins, the Scholar explains to the Master why he wants to learn about astronomy: "I see in the heaven marvailous motions, and in the reste of the worlde straunge transmutations, and therefore desire muche to knowe what the worlde is, and what are the principall partes of it, and also how all these straung sightes doo come" (3). The Master's response is to explain the basics of the Ptolemaic system and then to provide lessons in the construction and use of an armillary sphere, a three-dimensional model of the universe that would allow the Scholar to visualize the movements of the heavenly bodies from an earthly perspective. Late in the book the scholar asks specifically about the retrograde motions of the planets, and the Master mentions the Copernican theory that would destroy the model of the universe that most of the book has been devoted to explicating. The "marvailous motions" and "straunge sightes" that the book explains, then, seem to be astronomical phenomena rather than astrological predictions.

In this Copernican moment, issues of what can or cannot be seen become especially important. When the Master is describing the number of concentric spheres in the universe, he takes issue with those who had posited the existence of ninth and tenth spheres in order to explain precession of the equinoxes. He tells the Scholar that there is clearly an eighth sphere because "this heaven is manifest inough to all mennes eies, so that no man nedeth to doubte of it, for it is that skye wherein are all those starres that we see, except the five lesser planets" (7). The Master can still insist that the fixed stars "do stande and continue in one forme or distaunce eche from other, and chaunge not there places in their spere" (8), but he will not describe ninth and tenth spheres: "I thinke it best to tell you of no mo spheres, then are perceptible by sighte, for so manye are we certaine of" (10). The Master here emphasizes an instance where Copernican theory corresponded more closely with what could be actually observed, since he explained precession as caused by the movement of the Earth, doing away with the need for the invisible ninth and tenth spheres.

Perhaps because of the Master's insistence that heavenly phenomena need to be observed if their existence is to be credited, the Scholar is scornful of Copernican theory. When the Master explains the motions of the sun and moon according to the Ptolemaic system, the Scholar responds: "this maye all men see, that can see any thing. Yet have I heard of some so grossly witted, that they doubted which way the Son and Moone dyd come into the east agayne, as though they dyd not thinke that the skye dydde move about the earthe" (11). The Master at this point seems to agree: "such grosse ignorance happened sometymes to famous men, for lacke of due consideration of that, whiche all men maye see, as I will in place conveniente more largelye note" (11).

Fairly early on in the treatise, the Scholar asks the Master to explain "the appearaunces which are noted in the heavens" and "the manner of their motions" (15). The Master can only explain these motions with the use of a "Materiall sphere," the three-dimensional model of the geocentric universe that would allow the Scholar to visualize the movements of the heavenly bodies. The bulk of the treatise contains instructions for building such a sphere and an account of the movements of the sun and planets in relation to the signs of the zodiac, without touching on the complexities that were challenging the Ptolemaic universe. At one point, the Scholar remarks that whatever Ptolemy has argued must be correct: "I heare all learned men say, Ptolemye is the father of that arte, and proveth all his woordes by stronge and invincible reasons" (126). The Master responds that while Ptolemy is praiseworthy due to "his travell being so great, his diligence so exacte in observations" (127), the Scholar should not therefore accept everything he argues based solely on his reputation and authority: "yet muste you and all men take heed, that both in him and in all mennes workes, you be not abused by autoritye, but evermore attend to their reasons, and examine them well, ever regarding more what is saide, and how it is proved, then who saieth it, for autoritie often times deceaveth menne" (127).

At the end of the book, the Master adds a chapter that provides proofs for some of the phenomena that he has been describing, and he returns to the Scholar's difficult questions about precession and planetary motion. It is in this section that he explicitly mentions Copernicus, although he does not explain how Copernicus resolves these issues. As noted in chapter 2, the Master informs the Scholar that "Copernicus, a man of greate learninge, of muche experience, and of wondrefull diligence in observation, hathe renewed the opinion of Aristarchus Samius, and affirmeth that the earthe not only moveth circularlye about his owne centre, but also may be, yea and is, continually out of the precise centre of the world 38 hundreth thousand miles." The Master will not explain the issues at stake, however, because "the understanding of the controversy dependeth of profounder knowledg then in this Introduction may be uttered conveniently" (165). The Master here implicitly notes the threat posed by Copernicus to Aristotelian doctrine about the nature of the elements: the element earth by nature seeks the center. If the Earth is, in fact, 380,000 miles away from the center of the universe, then the whole system of sublunar physics is called into question. The Scholar, having been schooled by the Master in the importance of visual evidence in determining the number of spheres and still overly dependent on authority, protests, "I desire not to heare such vaine phantasies, so farre againste common reason, and repugnante to the consente of all the learned multitude of Wryters, and therefore lette it passe for ever, and a daye longer" (165). The Master gets the last word, which falls short of a clear endorsement of the Copernican theory: "you are too yonge to be a good judge in so great a matter: it passeth farre your learning" (165).

The Master uses the inability of the Scholar to understand complex matters as an excuse not to go into the vexed issues of precession and retrograde motion, both phenomena that made Ptolemaic theory less plausible and that were explained more satisfactorily by Copernicus. Returning again to the number of the heavenly spheres, the Scholar supports the idea that there must be ninth and tenth spheres to explain precession and trepidation: "if it be true that there be suche varieties of motions, then it seemeth reasonable to assigne so many spheres as there be motions severall" (278). The Master tries to blow him off again without a full explanation (which would presumably involve the Copernican argument that precession is explained by the movement of the earth): "although you thinke so now, you may be persuaded peradventure to thinke contrary hereafter, as most wise men in that arte do" (279). Recorde allows the Scholar to express impatience with this nonanswer: "but in the meane ceason what shall I thinke?" (279). Similarly, the Scholar presses for a better explanation of the "straunge motions" that he asked about at the beginning of the treatise, requesting an explanation of "the progression, retrogradation, and station of the Planetes" (279). The Master beats around the bush, admitting that the answer would be "obscure" and again urging the Scholar to be satisfied with the instruction he has received in basic Ptolemaic theory. The Master refuses to answer these hard questions; however, Recorde causes them to be raised, making clear that in this "meane ceason" of unsettled theory, it is not clear what people should think about the physical shape of the universe or whether they can rely on ordinary experience for information about it.

OPTICS AND THE COPERNICAN MOMENT

The 1560s to 1571 saw the publication of treatises that, in different ways, call on ideas from optics to bridge the increasing split between the manifest evidence provided by the world and the things that could not be explained in terms of that evidence. John Dee derived from various sources an ambitious theory that explained many difficult or inexplicable natural phenomena as caused by invisible rays, which could be controlled and focused by lenses. William Fulke published several treatises designed to provide naturalistic explanations for seemingly marvelous or supernatural meteorological phenomena, calling on phenomena of refraction and reflection to explain strange visions in the air. Thomas Digges provided evidence of his father's experiments with concave mirrors and other lenses to create "perspective glasses" that functioned like a telescope.

Although these writers used optics or "catoptrics" for different purposes, they all seem to have been drawn to theories that explained the ways in which vision could be manipulated, extended, and deceived. The Copernican universe essentially involves a change in perspective, taking the sun and fixed stars rather than the Earth as the "point of reference" from which to observe the universe. As E. A. Burtt has argued, "to Copernicus's mind the question was not one of truth or falsity, not, does the earth move?" Instead, he allowed for the possibility of a change in perspective and "simply included the earth in the question which Ptolemy had asked with reference to the celestial bodies; what motions should we attribute to the earth in order to obtain the simplest and most harmonious geometry of the heavens that will accord with the facts?"[30] Copernicus had argued in the *De revolutionibus* that some visible phenomena, such as the apparent retrograde motion of the planets, were in fact optical illusions caused by the motion of the Earth.[31] Kepler comments in his *Optics* that "as regards Copernicus, however, this whole illusion (*phantasia*) of standing still and retracing of steps is demonstrated most beautifully from optics."[32] In the 1560s, optics offered the possibility of resolving a number of perceptual issues and providing new mastery over nature.

The new ferment of ideas that caused Recorde simultaneously to insist on and deny the evidence of the senses, as well as to raise and refuse to consider the most problematic aspects of the Aristotelian universe, drove John Dee to espouse theories about the natural world that today seem quite strange. Yet, Dee's ideas are grounded in the same kinds of mathematical knowledge as his friends' and followers' and, rightly understood, are, at least in his *Propaedeumata* and in parts of his preface to Euclid, still largely within the realm of Aristotelian naturalism.[33] In the 1550s and 1560s, Dee was still optimistic that an ambitious program of observation and quantification could provide a rational grounding for astrology. He followed Roger Bacon in believing that many of the so-called "occult" phenomena in the Aristotelian universe were caused by "species" or invisible rays, which were given off by heavenly bodies and potentially by other material objects.[34] In his *Propaedeumata* of 1558, Dee uses the enigmatic language of the secrets tradition to delineate a program for quantifying the rays given off by heavenly bodies and therefore providing a rational basis for astrological prediction, as well as the basis for advances in medicine and optics. Dee believed that concave and convex mirrors could focus the stars' rays to great effect, as a magnifying glass can focus the sun's rays to start a fire. Rather than being disturbed, like Recorde, by phenomena that could not be observed, Dee felt that the study of these invisible rays could make the secrets of nature both manifest and subject to man's control.

In the *Propaedeumata*, Dee, unlike Recorde, explicitly aligns his text with the secrets tradition: "not only are those things to be said to exist which are plainly evident and known by their action in the natural world, but also those which, seminally present, as it were, in the hidden corners of nature, wise men can demonstrate to exist."[35] The next aphorism goes on to explain that these "actions" emit "spherically upon the various parts of the universe rays, which, in their own manner, fill the whole universe."[36] Dee argues that the effects of these rays are sometimes visible, sometimes invisible: "everything that exists in the world of elements, no matter how paltry, is an effect of the total celestial harmony or a particular example and reproduction of it. But this appears more clearly in some things than in others" (189). Dee calls on the archetypal example of an "occult" phenomenon to illustrate a visible example of the action of these rays: "in the magnet, God has offered to the eyes of mortals for observation qualities which in other objects he has left for discovery to the subtler research of the mind and a greater investigative industry" (133). The magnet, says Dee, has the "ability to penetrate solid bodies with its essential rays" (133).

For Dee, optical devices offer the possibility of controlling nature and of making its hidden secrets manifest. He argues that mirrors and lenses provide the possibility of focusing the rays of the stars: "if you were skilled in 'catoptrics,' you would be able, by art, to imprint the rays of any star much more strongly upon any matter subjected to it than nature does itself. This, indeed, was by far the largest part of the natural philosophy of the ancient wise men" (149). Dee expands on the role of optics in making secret knowledge visible in his preface to Billingsley's edition of Euclid in 1570.[37] He identifies "Archemastrie" as an "Arte" that "teacheth to bryng to actuall experience sensible, all worthy conclusions by all the Artes Mathematicall purposed, and by true Naturall Philosophie concluded" (A3r). He goes on to explain that the "chief Science, of the Archemaster, (in this world) as yet knowen, is an other (as it were) OPTICAL Science: whereof, the name shall be told (God willyng) when I shall have some, (more just) occasion, thereof, to discourse" (A3v). Dee notes that "the Astronomer, and the Opticall Mechanicien, put some thyngs in *Experience*: but neither, all, that they may: nor yet sufficiently" (A3v). Dee seems to hope that optical devices will someday allow invisible facts about the universe to become manifest—as, in fact, happened when Galileo used a telescope to confirm the Copernican hypothesis.

Dee's allusion to potential advances by "Opticall Mechaniciens" is probably referring to a reflecting telescope that was constructed (or claimed to have been constructed) by Leonard and Thomas Digges and described in their treatise of 1571, *A Geometrical Practise, named Pantometria*, which provides instruction in the application of geometry to measurement of land and mapping.[38] Following directions for using a mirror to measure the distance from shore of a ship, Digges inserts a digression describing the "marveylouse . . . conclusions that may be performed by glasses concave and convex of circulare and parabolicall fourmes" (G1v). He goes on to claim that "ye may by applycation of glasses in due proportion cause any peculiare house, or roume thereof dilate and shew it selfe in as ample fourme as the whole towne firste appeared, so that ye shall discerne any trifle, or reade any letter lying there open, especially if the sonne beames may come unto it, as plainly as if you wer corporally present, although it be as distante from you as farre as eye can discrye" (G2r). Digges further claims that his father was able to "read letters, numbered peeces of money with the very coyne and superscription thereof, cast by some of this freends of purpose uppon Downes in open fieldes, but also seven myles of[f] declared what hath beene doon at that instante in private places" (A3v).[39] He goes on to talk about the power of lenses to focus the rays of the sun "to fire, pouder, or any other combustible matter" (G2r). Whether or not these claims are exaggerated, they demonstrate that there was particular interest in the powers of lenses and mirrors among members of the Dee circle during this period.

William Fulke, a writer, Protestant apologist, and developer of board games, also called on catoptrics in his meteorological treatise of 1563, *A Goodly Gallerye*.[40] In some ways, Fulke could not have been more different from Dee. His first book, the *Antiprognosticon* of 1560, is, in the words of the subtitle, "an Invective against the vayne and unprofitable predictions of the Astrologians."[41] And yet, Fulke's attack on astrology derives from an impulse not all that different from what lies behind Dee's astrological program in the *Propaedeumata*. While Dee is concerned in that work with identifying and quantifying the natural properties of the stars within a framework that is largely compatible with the Aristotelian system, Fulke is perhaps more dedicated than anyone else in sixteenth-century England to explaining all phenomena in terms of Aristotelian naturalism.[42] The difference between them lies in Dee's emphasis on invisible rays or species, a concept derived from the "secrets" tradition and his belief that these rays mediated between natural and supernatural causes. Fulke, on the other hand, accepted God as the first cause of all events in the cosmos but believed that all heavenly and earthly phenomena had naturalistic explanations.

In his *Antiprognosticon*, Fulke goes beyond Leonard Digges in confining prognostication to phenomena that can be directly observed. Although his ultimate motivation for debunking astrology may be religious, coming as a response to predictions by Nostradamus that were an impediment to the progress of the reformation, Fulke's grounds for dismissing astrological predictions are firmly based in Aristotelian naturalism.[43] In this approach, he is not so far from Dee's argument in the *Propaedeumata*; both accept that the stars have power over earthly matter. In Fulke's words, "the starres have dominion and exercise their power on mens matters contingently, that is, as it were a certaine inclination or some secrete motion."[44] However, Fulke argues that these effects are too complicated and too easily disrupted by other phenomena to yield certain predictions: "as for the signification of the starres both we learne by experience, neither do the Prognosticators denye, but that they may, by divers meanes be averted or tourned to some other effect" (B8v). Dee would agree but differs from Fulke in offering an ambitious program to focus and quantify these muddled and complex effects.

Fulke offers an extended discussion of Genesis 1:14, insisting that the "signs" and "tokens" mentioned in the Bible are simply signs of natural phenomena themselves. He therefore directly articulates the view that nature offers only signs of its own workings: the signs mentioned in Genesis 1:14 are not "signes . . . for ye purpose of their predictions" but instead "they shalbe tokens of spring, somer, autume, and winter, of dais natural and artificial, long or short of yeres, according to ye sons course, or to ye moones course" (B5v–B6r). Thus the sun and moon offer "Signes of hete when ye son draweth nere to us, of cold when he departeth, of abundance of humors when ye moon encreaseth, and ye contrary when she is in the wane or decreasing" (B6r). Thus, "if the Mariner by his signes can forshewe the tempest to folowe, the Phisition by his tokens, can forsece the increace or declination of the siknes" these have "sure groundes" which astrology lacks (C7r–v).

For Fulke, astrology has no scientific basis because it cannot be observed or experienced, and he remains firmly within the Aristotelian epistemology of manifest qualities. He argues that "this is common to all sciences, that they may bee demonstrated" so that even if they "be of suche nature, that they canne not be shewed and confyrmed by things more general . . . yet by demonstration or induction they maye be so plainly sette before our eies, that no man neede to doubte, that they are moste true and certain" (B2r–B2v). Fulke imagines that astrologers will object that the distance of the stars means that their effects cannot be made manifest: "although this is easyly to be shewed in such thynges as almost we may perceive by our sences, yet the same in thynges farther of; can scarsly be declared, as ye greatnes of the stares, or their distance from the earth, or from themselves one from another" (B3r).

Bauckham has argued that Fulke is the author of a manuscript that mentions, and discounts, Copernican theory on the grounds that it could not explain the visible phenomena that the Aristotelian system could account for: "I colde . . . yelde my consent, if the accesse and recesse of the first movable, the alterations of the severall latitudes of the planetes, the continuall course and recurse of the sea, and blowing of the wind in all quarters of the world wold suffer me."[45] Fulke places such emphasis on what could be observed through daily experience that he can't accept a theory that doesn't address these observations. If this manuscript was written by Fulke, it suggests a Copernican context for his insistence on daily observed experience and his appeal to optics to explain a universe whose appearances were threatened.

Fulke, like Dee, calls on scientific technology as a method for making things that "scarsly can be declared" manifest to human senses. In this case Fulke refers to techniques for using the angular size of the shadow of the Earth during a lunar eclipse to compute the size of the moon and its distance from the Earth. He argues that although computing the size and distance of the heavenly bodies "be difficult, and suche as can hardly be persuaded to the common sort: yet as difficult as they be, and harde to knowe, there is a meane whereby men maye come to the science and understandyng of them" (B3r). Fulke appears to be one of the first writers to use the secrets tradition language of travel to an inaccessible place to describe a mathematical or scientific technique: "we have the shadowe of the earthe, as it were a ladder whereby we ascende into heaven, and beholde many thynges, whereunto grosse wittes cannot frame. But no waie is it possible, that the principles of this arte of Astrologie, may be either demonstrated or proved" (B2v).

Bauckham has called Fulke's *Goodly Gallerye* "the only properly scientific discussion of the subject [of meteorology] in English in the sixteenth century."[46] Like

Leonard Digges, Fulke accepted as fact the accounts of marvelous atmospheric phenomena such as visions of armies in the sky, the appearance of multiple suns and moons, and so on, but unlike Digges, who distanced himself from these accounts by carefully citing their sources, Fulke provided naturalistic explanations for them. In the *Goodly Gallerye*, for instance, he argues that the "signification" of comets (which he explains in the usual way as exhalations that have been drawn up into the atmosphere and set on fire) have natural causes: "They ar sayd to betoken drought, barynes of ye earth, and pestilence," "drought because a *comet* can not be generated without great heat, and muche moisture is consumed in ye burning of it"; "Barrines, because ye fatness of ye earth, is drawen up, whereof the *comet* consisteth"; and pestilence because the comet "corrupteth the ayre" (15v–16r). He considers the "civile or politike effects" (15v) of comets to be more of a reach but ventures the idea that comets appear when there are "many whote and drie *Exhalations* in the aire, which in drie men kindle heat, whereby they ar provoked to anger, of anger cometh brawling, of brawling fighting and war, of warre victory, of victory chaunge of common welths" (16r–16v).

Fulke calls on optics to explain more difficult phenomena. Noting that "oftentimes, men have seen, as they thought in the firmament, not only two sonnes, but oftener thre sunnes, and many more in nomber" (40v), this phenomenon, though "wonderful" occurs when clouds act like a mirror: "they are nothing els but Idols, or Images of the sunne, represented in an equal smooth and watry cloude, placed on the side of the sunne, and sometime on both sydes, into which the sunn beames being received as in a glasse, expresse ye likenes of fashion and light, that is in the sunne, appearing as though there were many sunnes, where as in dede there is but one, and all the rest are images" (40v). He offers a more elaborate optical explanation for the appearance of crosses, circles, armies, beasts, and other images in the sky. These apparitions "perteyne to the knowledge of *Optice* and *Catroptice*, that teache howe by diverse refractions and reflections of beames, such visions are caused" (45r). Images of armies and monsters in the sky can be caused by reflections in two ways "the one artificially, the other naturally" (45v). Artificial apparitions are caused by "certein glasses and instruments made according to a secret part of that knowledge which is called *Catoptrice*" (45v–46r). Fulke is evidently imagining something like a giant, aerial slide projector. Most such visions have natural causes, however, "when the disposition of the ayre, hath been suche, that it hath received the image of manye thinges placed and done on the earth" (46r). While Dee hopes that optical devices will reveal the secrets of the heavens, Fulke sees the sky itself as an optical device creating effects that can not otherwise be explained according to Aristotelian naturalism. Dee believes that optics can make secrets manifest; Fulke believes that optics reveals seeming secrets to be illusory.

I have been arguing that in the wake of the dissemination of Copernican theory in England, educated writers who were aware of and understood the theory were presented with the possibility that the universe that was made manifest through daily experience was not as it appeared to be. This possibility threatened the foundations of Aristotelian epistemology. Writers responded to this threat with a closer scrutiny of the relationship between nature's visible signs and its causal reality, and they became interested in optics as a way to account for (or master) the growing gap between the evidence of the senses and the operations of nature. While these midcentury writers were variously acknowledging and explaining away the Copernican perspective on the movements of the Earth, sun, moon, and planets, they clung, perhaps more fiercely than had astronomers in the past, to a feature of the Ptolemaic universe that Copernicus had preserved: the fact that the fixed stars were unmoving and unchanging.

Robert Recorde's Master (in the *Castle of Knowledge*) informs the Scholar that the fixed stars "do stande and continue in one forme of distaunce eche from other, and chaunge not their places in their spere" (8). Recall that the Master debunks the idea that there are ninth and tenth spheres, added to account for the slow movement of the fixed stars over thousands of years and the resulting shift of the signs of the zodiac; he insists that the eighth sphere "is manifest enough to all mennes eies, so that no man needeth to doubt of it," while those additional spheres are not manifest and therefore unlikely to exist. When the Scholar insists that additional spheres are the only way to explain "such varieties of motions" on the part of the fixed stars, the Master explains that once the Scholar learns more he will be "persuaded to think the contrary" (179), possibly alluding to the fact that the Copernican theory explained the apparent motion of the fixed stars as a function of the uneven rotation of the earth.

For Dee the stability of the fixed stars becomes even more important, perhaps in part because his program of quantifying stellar influences already has so many variables that if the stars were capable of movement, his goal would be infinitely more difficult. In the *Propaedeumata*, Dee asserts, "that the mutual spaces among the fixed stars have never been altered in the whole eternity of time shows that the stars are much superior even to those things in the elemental universe that strongly retain an unvarying condition in their situations" (161). For Dee, the slow movement of precession explains why "changes and vicissitudes occur even in our most important affairs, or in those of which we not judge the most stable and self-consistent," perhaps explaining why astrology is so difficult to render an exact science: "it is necessary that the nature of the event which is performed or specifically signified by two fixed stars, or by many, should be diverse" (163). The fixity of the stars is crucial to the stability of the cosmos, for without it, "no particle would be preserved naturally even for a single day" (161). Similarly, Fulke (in the *Goodly Gallerye*) takes pains to rebut the Epicurean argument that "shotyng and falling Starres" are actual "starres" that "fall out of the firmament" (7v–8r). This cannot be the case because "God hath set them fast for ever" (8r).

THE NEW STAR OF 1572 AND THE NEED FOR SECRET KNOWLEDGE

The unchanging nature of the fixed stars, then, remained a touchstone in a cosmos that in other ways began to appear unstable and unintelligible. In November 1572, however, this belief in the fixity and stability of the firmament was threatened when a new star appeared in the constellation Cassiopeia. This supernova was impossible to ignore, as it "shone brighter than Venus at its maximum brilliance, then gradually died out until, nearly seventeen months later, it became once more invisible to the naked eye."[47] Francis Johnson explains that "the research and discoveries in connection with this famous new star did more than anything else to bring about the downfall of the Aristotelian cosmology."[48] Because the star was so large and bright, several astronomers were able to see that there was no observable parallax, proving that the new object must be located above the moon in the firmament of fixed stars where change was supposed to be impossible. The appearance of the new star seems to have revived interest in the phenomenon of precession, which also appeared to provide evidence that the fixed stars had changed positions over time (unless a ninth and tenth spheres were posited to make this movement part of the accepted system). That a new star could appear, and then disappear, brought the changelessness of the firmament into question, at least for some people, but there was a range of reactions to the new star.

Deborah Harkness has argued that the appearance of the new star in 1572, combined with the appearance of a comet in 1577, the earthquake in 1580, and a significant astrological event (a "grand conjunction") predicted for 1583, caused a period of "heightened eschatological fervor."[49] She places Dee's attempts to communicate with angels in this context, suggesting that he had become increasingly frustrated by his inability to make progress in unlocking the secrets of nature and felt that these signs suggesting the approaching end of the world made his need for knowledge more acute.[50] While Dee was an extreme case, Thomas Digges, who wrote treatises on the new star and the Copernican universe during this period, also seems to have been increasingly influenced by Neoplatonism and the secrets tradition. For both Dee and Digges, the ideas derived from various esoteric traditions provided ways to make the unsettling of the Aristotelian cosmos more intelligible and less threatening. Other writers in this period between 1572 and the end of the century found other ways to deal with the shocking fact of the new star; most, however, retain a grounding in Aristotelian naturalism while appealing in different degrees to secret sources of knowledge.

After 1572, the appeal to optics as a way to see the truth about the universe gives way to increased use of the trope of travel to previously inaccessible realms.[51] The supernova seems to have ended the hope that the unsettling of the cosmos was an optical illusion, or that a combination of mirrors and lenses could provide access to some stabilizing truth. Deborah Harkness relates Dee's angelic conversations to the biblical image of Jacob's ladder in Genesis 28:12, where Jacob dreams of a ladder that ascends from Earth to heaven and sees "the angels of God ascending and descending on it."[52] Authors in this period expand on this image to depict a range of techniques as providing a way for an observer to travel up to the heavens, into the depths of the earth, or into the human body. This new need to somehow see the secrets of nature at first hand reflects, as Harkness suggests, increasing fears that the "book of nature" was becoming unintelligible in its old age.[53] There is less confidence that "signs and tokens" offer access to causal truth or that human observers can interpret them without help—help that now needs to be powerful enough to bridge the gap between appearance and reality.

It's not a surprise, then, that Thomas Digges's Latin treatise on the supernova of 1572 is entitled *Alae seu scalae mathematicae*. A translation of the full title is "Mathematical Wings or Ladders, with which to ascend to the farthest Theatres of the Heavens and with a new and unheard-of Method, to explore the paths of all the Planets, and then to find out the Distance, Position and immense magnitude of that portentous Heavenly Body shining with remarkable brightness in the Northern part of the world."[54] The "wings or ladders" offered by the treatise were "the advanced spherical geometry developed in the *Alae*, and the superior observations made possible by better instruments," resulting in a more accurate measurement of parallax, allowing more accurate determination of the distance from the earth of the heavenly bodies.[55] Digges promises the reader that he is offering "no vulgar stairs in this little mathematical book, by which you will be able to ascend the highest peaks of the sky and to tread with your swift foot the seats [of heaven]; nor is it marvelous that any human ingenuity can penetrate to the highest threshold of the sky and bring back the treasures of God to mortals" (A1r).

Despite his references to "Platonic wings," Digges remains pretty firmly within the fold of Aristotelian naturalism and, despite the evidence of the new star, maintains the distinction between a mutable "elementary" realm and an immutable celestial realm. While some observers argue that the new object in the sky is a comet, Digges insists that calculation of its parallax will reveal that it is located "far off from the comets with respect to its location, since in the very purest Aether itself there are not able to exist in accordance with Nature any changes, mutations, or alterations, to which the vapors of comets are not able to be carried by any force (as Aristotle says, with all Physicists consenting)."[56]

On the other hand, he believes that the most likely explanation for the appearance of a new star is supernatural: like the star that appeared to the magi at Christ's birth, this new star has been sent by God to herald the second coming. It is therefore of crucial importance that he prove that the object is not a comet or ordinary star so that people will not "either have contempt for this very rare sign and true miracle and messenger of God, or else think it isn't a big deal."[57] At the same time, Digges believes that accurate observation of parallax provides the key to determining whether the Ptolemaic or Copernican theories are true: "whether the Earth lies quiet and immovable at the center of the World, and whether that huge mass of moving and fixed Orbs rotates in a circle by a most rapid course in the space of 24 hours, or rather, that that immense sphere of fixed stars remains truly fixed and that apparent motion occurs only from the circular rotation of the Earth with reference to the celestial poles themselves."[58] The wings or ladder of parallax, then, simultaneously prove that the new star is located in the realm of the fixed stars but also proves the Copernican hypothesis that it is the Earth, and not the fixed stars, that moves. If Copernicus was right and the stars do not move even in precession or trepidation, then the appearance of something new in the firmament can only be a miracle of God, which, can in turn only be verified as such by scientific observations.

In his appendix to his father's *Prognostication* (1576), Digges provides a clearer sense of the relationship between human observation and the truth about the universe, accepting the fallibility of the senses as the price to be paid for confirmation of the stability of a universe anchored by the immobile fixed stars. In this text, both Neoplatonism and the metaphysical optics of Roger Bacon seem crucial to Digges's acceptance of a heliocentric system where "the whole Globe of mortalitie is caryed yearely around the Sunne, which like a king in the middest of all raygneth and giveth laws of motion to the rest, sphaerically dispersing his glorious beames of light through all this Coelestiall Temple" (M1r). Digges here seems to echo Dee (who follows Bacon) in emphasizing the rays of the sun dispersed "sphaerically" throughout the universe.

Digges seems to see as the major benefit of the Copernican system its solution to the problem of precession. He acknowledges that the new system depends on a change in perspective, since Copernicus has taken "the immovable sphere of fixed starres, containing it selfe and all the rest, and therefore fixed: as the place universall of rest, whereunto the motions and positions of all inferiour spheres are to be compared" (M4v). This change in perspective simultaneously preserves the fixity of the stars and also confirms the evidence of the senses that there are no spheres beyond the eighth: "For albeit sundrie Astrologians finding alterations in the declination and longitude of Starres, have thought the same also should have his motion peculiar: yet Copernicus by the motions of the Earth salveth all, and utterly cutteth off the ninth and tenth spheres, whiche contrarie to all sence the maintainers of the earths stabilitie have been compelled to imagine" (N1r).

For Digges, then, securing the stability of the fixed stars is worth acknowledging that the manifest evidence of the senses—our earthly perspective that we are still and everything else moves—is wrong. Digges easily affirms the movement of and the deceptive nature of our ordinary experience of the Earth, which, "having his peculiar and strange courses turning every 24 houres rounde upon his owne Centre: whereby the Sunne and the great Globe of fixed starres seems to sway about and turne, albeit indeed they remayn fixed. So many wayes is the sense of mortall men abused" (M1r). Here again, something like Neoplatonism helps Digges to separate the evidence provided by reason from the evidence of the senses, comparing reason to a central illuminating light like the sun: "Why shall we so much dote on the appearance of our sences, which many wayes may bee abused, and not suffer ourselves to be directed by the Rule of Reason, which the great GOD hath given us as a lamp to lighten the darknes of our understanding, and the perfect guide to lead us to the golden branch of Veritie amidde the Forrest of errours" (M1v). Reason in this case needs the help of mathematical knowledge, although Digges does not use the language of wings or stairs here, suggesting that the light provided by the sun is sufficient to provide accurate knowledge: "and all these things, although they seem hard, strange, incredible, yet to any reasonable man that hath his understanding ripened with Mathematicall demonstration, Copernicus in his Revolutions according to his promise hath made them more evident and cleare then the Sunne beames" (M4v).

As we have seen, Francis Shakelton's *A Blazyng Starre* of 1580 provides evidence that Digges's learned Latin discussion of the new star was read with at least partial understanding by people who had little mathematical training. Shakelton's treatise views the new star and rash of comets in the 1570s as signs sent by God to warn humanity to repent before the imminent end of the world. Shakelton clearly differentiates his own approach and method from that of "some right skilfull in the Mathematiques."[59] He acknowledges that his treatise has different rhetorical aims and argumentative methods than that of Digges, noting that his own "profession, drawying me another waie, requireth an other maner of methode" (a3v).

Shakelton understands and embraces Digges's argument that the star of 1572 was "no Comete, but mere supernaturall, directly apposite to the Starre whiche appeared at the Nativitie of our Saviour Christe" (D3v). Because the star is supernatural in origin, it "maie worthely be tearmed or named a signe, unto the whole yearth. As more at large appeareth in a Booke, which was written by that worshipfull gentleman maister Thomas Digges, called *Alae seu scalae mathematicae*, where the miracle is by demonstrations mathematicall, plainly and manifestly proved" (D3v). As did Digges himself, Shakelton sees the data about parallax as providing mathematical proof of the supernatural origins of the star.

The purpose of Shakelton's treatise is explicitly eschatological, arguing that God has sent "this burning Comete, whiche now appeareth: as also, by the late dreadfull yearthquake, but most cheefely by that supernaturall Starre of all others, that was seene in Cassiopeia . . . to move and stire up all the sorte of us (beeying moste wretched and miserable sinners) to be mindfull of the Judgement daie" (D6r). Nevertheless, his understanding of the world is grounded in Aristotelian naturalism and he applies criteria similar to those used by Leonard Digges and Fulke to parse the complex causes of seemingly miraculous events. He calls on observation and experience of the manifest elements to support his argument that nature is experiencing senescent decline: "doe we not see the yearth to be changed and corupted: Somtymes by the inundation of waters? Sometymes by fiers? And by the heate of the sunne? And do we not see that some partes of the same ware old, and weare awaie even for verie age?" (A4r). He also separates out the "naturall causes" of "blasyng Starres"—which are, according to "Aristotle and divers other Philosophers . . . but an yearthly vapour, somewhat thicke, and clammie, and fatte in substaunce," which is drawn up to the sphere of fire and ignited—from their supernatural cause, being made by "the Lorde God hymself" who is their "original authour" (C4v–C5r). Like Leonard Digges, he questions the reports of armies seen fighting in the sky in Germany in 1547 because "wee can not render a naturall reason thereof" (C6r).

Shakelton does not seem to grasp the implications of Digges's argument that parallax can confirm the Copernican model of the universe. For Shakelton, "the earth (which is the centre of the worlde)" (A4v) has not been shaken from its place. He cites "Mathematiques" who argue that "ye constitution of the celestiall worlde, is not the same that it hath been in tymes paste" (A4v), viewing new calculations of the distance of the sun from the Earth as one more sign of the decline and imminent end of the cosmos. Deborah Harkness argues that Shakelton's views about the aging and decline of nature are similar to those of Dee in this period.[60] Shakelton, however, turned to orthodox religious belief as the solution, while Dee stretched the boundaries of science and religion by consulting angels. However, both inhabit an epistemological space that is recognizably based in Aristotelian naturalism and manifest experience of the world aided by mathematics, with an appeal to the supernatural for help in understanding the remaining secrets of a declining universe.

THE GROWTH OF SECRET KNOWLEDGE, 1580 TO 1600

The last twenty years of the sixteenth century saw an intensification of appeal to various "secrets" traditions as the explanatory power of Aristotelian naturalism continued to erode in the wake of Copernicanism and the supernova. Although scholars now see him as less influential than did Frances Yates, Giordano Bruno on his visit to England in 1583–85 did introduce an idiosyncratic mix of Copernican, Neoplatonic, Hermetic, and atomistic theories to a range of people, including members of the Sidney circle.[61] A group of thinkers associated with the Earl of Northumberland, among them Nicholas Hill, William Warner, and especially Thomas Harriot, pursued interests in alchemy, mathematics, and optics that led them to consider theories of atomism. The first extended Paracelsan medical treatise in English was published by Richard Bostocke during this period, and it sought to replace Galenic medicine with doctrines inspired by the alchemical and astrological medical theories of the Swiss writer Paracelsus.

In the 1580s and 1590s there was a shift from expecting that ordinary experience would reveal the truth about nature to sensing that many important truths are hidden and need some kind of special practice to be retrieved. Spreading knowledge of Copernican theory raised various epistemological possibilities. Copernicus's arguments could be seen as a way of resolving the problem of precession and restoring the fixity of the fixed stars without needing the (invisible) ninth and tenth spheres, that is, of making the phenomenon congruent with what could be observed without giving up the immutability of the firmament. Some, like Richard Bostocke and Gabriel Harvey, seem to have embraced Copernicanism as a solution to the problem of precession without understanding that it also entailed a heliocentric universe and a movable Earth. Copernicanism was acceptable to someone like Thomas Digges because, assuming that the Earth was necessarily imperfect and mutable, it was better to admit that the Earth moved (in defiance of experience) so as to maintain that the stars were fixed. Copernicanism could be seen as validating a Neoplatonic universe with the sun fixed in the center and the mutable and imperfect Earth revolving around it.

For those who understood that Copernican theory involved the movement of the Earth, its new truth was hidden, since ordinary experience didn't validate it. If the Earth was not at the center of the universe, then Aristotle's theory of the elements and their natural motions was evidently not correct. So it began to seem as if esoteric practices were needed to reveal the truth of nature both above and below the moon. At the same time, the new star, various comets, and the earthquake intensified the sense that the universe was seriously disturbed—and that these new practices were urgently needed to understand what it was trying to tell us.

Atomism emerged during this period as a way to explain material phenomena that Aristotelian theory struggled with: condensation and rarefaction, the continuity of magnitude, the simultaneous reflection and refraction of light through certain media. Although some historians of science have argued that the atomism of Harriot and his circle was derived from the writings of ancient atomists like Lucretius, I believe that John Henry is correct in arguing that Harriot's atomism, like Galileo's, was formulated in response to problematic phenomena and was not a fully worked-out theory based on Democritean atomism.[62] Harriot's willingness to hypothesize that matter was composed of invisible particles separated by invisible void spaces (rather than the continuous Aristotelian elements) may have been facilitated by this post-Copernican moment, when it became easier to accept a theory of matter not based in manifest qualities. Paracelsan theory also tended toward something like atomism in its conception of "seeds" that formed the basis of matter and that could both cause and cure disease, thus providing a more plausible theory of communicable disease than Galenic medicine could offer.

Secrets practices coexisted with naturalism; most people still continued to believe in basic meteorology and maintained the distinction between the sublunar and superlunary spheres. Anti-Aristotelian polemic increased at this point, but it also coexisted with continuing acceptance of many tenets of Aristotelian naturalism. The trope of travel to a previously inaccessible realm intensified—almost every writer in this period used it. The language of signs and tokens decreased—the universe was no longer expected to present signs that could easily be read. Its secrets needed to be pried out, or a difficult journey needed to be undertaken to find them out.

BOSTOCKE, HARRIOT, AND GILBERT: JOURNEY TO HIDDEN PLACES

In this period at the end of the sixteenth century, it is especially important to resist the temptation to classify thinkers and their ideas according to what they "got right" and what they "got wrong." Stephen Clucas has criticized those who want to see an "occult Harriot" as well as those who want to see a modern "progressive 'scientific' Harriot" and who see these elements in his thought as separable. These historians, Clucas argues, "are guilty of an 'over-differentiation' of Renaissance practices."[63] English writers during the last twenty years of the sixteenth century used a complex and contradictory mix of philosophies and methodologies to make sense of a world that suddenly seemed less intelligible: Aristotelian naturalism, Neoplatonism, atomism, Copernicanism, and advancing mathematics were all deployed without clear differentiation. Bruno accepted Copernicanism, atomism, and an infinite universe but for all the wrong reasons. Gilbert went farther than anyone in advocating an experimental empiricism and debunking questionable authority, but he accepted Aristotelian principles and had quasi-Hermetic ideas about magnetism as an animating force. All of these very different writers struggled with similar issues and used similar language to figure out their relation to a universe that seemed, on the one hand, suddenly unintelligible, and on the other, to be on the verge of yielding up its secrets.

Richard Bostocke's *The Difference between the auncient phisick* of 1585, William Gilbert's *De Magnete* of 1600, and Thomas Harriot's letters to Kepler in 1606–7 (which probably describe work done earlier) seem at first glance to have little in common. Bostocke's book is the first English text to provide an extended account of Paracelsan medicine and the natural philosophy behind it.[64] Gilbert's book sets out experimental results providing a (mostly) naturalistic explanation for the "occult" phenomenon of magnetism, and Harriot's letters respond to questions raised by Kepler about his experiments in optics. All provide good examples of the epistemological moment when the manifest qualities of the Aristotelian universe were being replaced by special knowledges of various kinds.

Bostocke's book provides an early example of the increasingly heated anti-Aristotelian polemic that would be taken up by later figures like Boyle.[65] Bostocke's title says it all: *The difference betwene the auncient Phisicke, first taught by the godly forefathers, consisting in unitie peace and concord: and the latter Phisicke proceeding from Idolaters, Ethnickes, and Heathen: as Gallen, and such other consisting in dualitie, discorde, and contrarietie. And wherein the natural Philosophie of Aristotle doth differ from the trueth of Gods worde, and is injurious to Christianitie and sounde doctrine*.[66] In a passage that seems directed against someone like Fulke, Bostocke objects to Aristotelian naturalism precisely because of its naturalism, arguing that it does not "admit any Metaphisicall principle in natural thinges" and thus "teacheth men to cleave and sticke fast unto the nature of thinges, not ascending nor considering the Creator. And so it tyeth thee (O God) to the second causes" (4r–v).

The book is as much religious tract as Paracelsan treatise. Bostocke avoids the common trope of travel to another realm because he is at pains to emphasize that God alone sees fully, although he has given mankind some access to his secrets through the practice of distillation. Bostocke sounds a bit like Thomas Digges when he describes "God which sitteth in the Center of all thyngs that be, from whence with his universall eye, that is with the light of his substance, whereby he is their being and life, doeth behold all things" (B3r). Just as Digges described the sun as lighting the world from the center and providing the illumination of reason to mankind, Bostocke describes God shedding light on the whole universe as he beholds it. However, for Digges, the sun helps man understand that a heliocentric universe coincides with reason if not with sensory experience. For Bostocke, it is God alone who has a clear view of the universe, and it is not important that he communicate it fully to man.

It may be that Bostocke's religious beliefs made him willing to give up Aristotelian positions that others clung to. After making the usual religious objections to the Aristotelian doctrines of the eternity of the world and the impossibility of creation ex nihilo (1r), he objects to the concept of fixed stars for the same reasons: Aristotle "very absurdly teacheth, that all the Orbes, be bodies subject to no corruption, and that they and their motions be eternall" (2r). He questions the differentiation between the areas above and below the moon on the same grounds: "it teacheth that God medleth not under the Moone, and that he is not the maker nor the creatour of any thing, but onely the mover of the heaven" (2v).

Bostocke is also willing to abandon the Aristotelian concept of manifest qualities in favor of secret knowledge. Aristotle is wrong because he "admitteth nothing, that cannot be demonstrated" (A5r). Bostocke explains that Galenic medicine fails because it "doth attribute the cause of things to dead qualities, of heate, cold, etc. and not to the lively vertues and powers in thinges" (A6v). For Paracelsan practitioners, "diseases doe proceede of the mechanicall spirites and tinctures of impure seedes joined to the pure by thy curse, O just God" (A6v). These "seeds" can function something like a germ theory of disease, although the Paracelsan theory of seeds was complicated and changed over time.[67]

For Bostocke, the humoral symptoms of disease are still signs and tokens, but he places a new emphasis on the secondary nature of their relation to truth. Aristotelian "humours and qualities" are "only dead accidents," and "conditions, signes, tokens, and as it were only flowers and colours of diseases, and not the very matter, cause, substance, or nature of the disease, they are caused and not the causes" (C6v). Instead of the Galenic practice of diagnosis through observation of humoral signs, Bostocke calls on the secrets tradition, seeming to advocate a mishmash of all the possible adept practices. He recommends "the true and auncient phisike which consisteth in the searching out of the secretes of Nature, whose study and use doth flowe out of the Fountaines of Nature, and is collected out of the Mathematicall and supernaturall precepts, the exercise whereof is Mechanicall, and to be accomplished with labour," variously called "Cabala," "sacra scientia," "Chymia," and "Alchimia," but actually, for Bostocke, boiling down (so to speak) to the practice of distillation: "by labour by the fire" (B1r).

Like other writers in this period, Bostocke substitutes for manifest experience an adept practice—in this case distillation—which provides a different kind of "visible and palpable experience" (D4v). The distiller "in trial of the nature of thinges, first spoyleth them of all their outward forms," using "the fire artificiall to purge and cleanse them" until there are only their "*Arcana,* inward and secret things remayning for the phisition to judge upon" (D6v). Bostocke goes farther than most writers in this period in explicitly denigrating what is manifest in nature in favor of what is secret: "the practicers in this arte [distillation] doe finde by experience by healpe of the fire, that eche thing hath two natures, that is, *occultum* and *manifestum,* and that *manifestum* is commonly contrarie to *occultum*" (D7r). The chemical physician teaches how "signes and tokens of diseases, be founded in the properties of nature: their originals be by them rightly discerned" (E6r). For Bostocke, "signs and tokens" are now denigrated to secondary and indirect indicators of causal phenomena. Paradoxically, he argues that "visible and palpable experience" (D4v) is provided not by ordinary observation but by the arcane practice of distillation, "so that the true proofe and tryal shal appeare to his eyes and [be] touched with his hands" (D5r). Because nature hides what is most important, true "experience" can only be obtained by boiling away the outer surface of things to reveal what lies within.

Because Thomas Harriot published virtually none of his work in mathematics, astronomy, optics, and chemistry, historians of science have disagreed about the nature of his contribution to the progress of the scientific revolution. Stephen Clucas traces various accounts of Harriot, from the old arguments by Muriel Bradbrook and Frances Yates that emphasized his participation in "occult" (in the old sense) circles, to Robert Kargon's study of his atomism, to arguments by other historians that Harriot rivaled Galileo in astronomy and Brahe in optics.[68] Hilary Gatti has shown that Harriot read and annotated works of Aristotelian naturalism such as Alessandro Piccolomini's commentary on Aristotle's *Meteorology*.[69]

I want to look briefly at a correspondence between Harriot and Johannes Kepler that took place from 1606 to 1608. The experiments that Harriot describes in the letters were probably carried out earlier, during the latter years of the sixteenth century. In these letters Harriot explains experiments in optics that led him to entertain an atomic theory of matter, and he defends that theory against Kepler's insistence on Aristotelian qualities. Although Harriot's table of experimental data, his rejection of Aristotelian qualities, and his tentative theory of atoms and vacua have been taken as precursors of modern science, he nevertheless uses the language of secrets, mysteries, and travel to a new place to describe his project.

Harriot's experiments involved shining a beam of light into various liquid media and observing the difference between the angle of incidence (the angle at which the beam of light strikes the liquid) and the angle of refraction (the angle at which the beam is bent by entering the denser medium). Kepler wrote to Harriot asking for information about his work in optics, and Harriot responded with a table of his observations and speculation about why light passing through a translucent medium was refracted and reflected. Harriot believed that an atomic theory of matter could explain this phenomenon even though matter manifestly appeared to be continuous and not composed of atoms and void space: "A dense diaphanous body, therefore, which to the sense appears to be continuous in all parts, is not actually continuous. But it has corporeal parts which resist the rays, and incorporeal parts vacua which the rays penetrate. So that refraction is nothing else than an internal reflection, and the part of the rays which are received inside, although to the sense it appears straight, is nevertheless composed of many straight line segments."[70]

Harriot here twice notes that the small "corporeal parts" and "incorporeal parts" that he posits and their actions in the liquid medium are at odds with the evidence of the senses; the water appears to the senses to be continuous, as Aristotle had argued it must be. However, Harriot's experiment forces the water to reveal that it is actually—"revera"—composed of invisible particles and void space.[71] Harriot therefore emphasizes that the material reality of the liquid is at odds with its appearance.

As he makes these arguments to Kepler, Harriot uses the language of travel to hidden realms, in this case *via* mathematics and the atom: "I have now led you to the doors of the house of nature where its secrets lie hidden. If you are not able to enter on account of their narrowness, then abstract and contract yourself mathematically into an atom and you will easily enter. And afterwards when you have come out tell me what marvels you have seen."[72] Harriot imagines that access to the secrets of nature involves mathematical abstraction from its concrete qualities and then a further act of contraction to the size of a tiny, invisible atom. The truth about nature is thus doubly removed from its surface appearance.

Kepler responds that he "was amazed by many things in the tables, which I had been ignorant of before."[73] He seems to accept the idea that Harriot has provided access to a previously hidden realm: "I think you have extended to me a key for seeking into the hidden natures of bodies."[74] However, Kepler refuses to accept Harriot's theory of atoms and void space, responding that Aristotelian qualities of "transparency and opacity" inhering within the same medium explain the coexistence of refraction and reflection.[75] In a letter of July 13, 1608, Harriot responds to Kepler that "if those assumptions and reasons satisfy you, I am amazed," reiterating that "I do not understand any transparency unless caused by a vacancy" or void space between atoms.[76] Harriot first refuses to expand on this theory of "the doctrine of the vacuum" because "I cannot philosophize freely," but at the end of the letter he can't resist providing one further argument. Although gold is "among our elements in composition held to be most dense and opaque," if you hammer it out to thin gold leaf and shine a light through it, "the light in its proper form can be seen, although it is green in color."[77] Harriot here interestingly anticipates the "gold foil" experiment carried out in the laboratory of Ernest Rutherford in 1909, which used a beam directed through thin gold foil to determine that atoms structurally consist of particles separated by empty space.

With William Gilbert's *De Magnete* we finally come to a work that has been recognized as making an original contribution to the development of the "new science."[78] The nature of Gilbert's achievement has been controversial, however, with some historians of science describing him as "the first experimental philosopher" and others viewing the *De Magnete* as "a backward-looking hodgepodge of Peripatetic and animistic philosophies of nature."[79] Peter Dear places Gilbert among proponents of "craft knowledge" since he used the writings of sailors and other practitioners, yet Dear acknowledges that Gilbert does not "present himself as one of their number, or indeed as any kind of mathematical practitioner. Instead he claims to be in search of *causes*: Gilbert strives to be a *philosopher*."[80] Elizabeth Spiller has argued that Gilbert works within a tradition of artistic "world making."[81] On the one hand, Gilbert rejected many previously authoritative accounts of the magnet and insisted on founding his account of magnetism on observation and experiment only. On the other hand, he retains a number of Aristotelian terms and concepts, was influenced by Paracelsanism, and comes close to describing the magnetically charged Earth as alive.

The late sixteenth-century natural philosophical writings that I have been tracing here provide a context that makes Gilbert's *De Magnete* seem less anomalous in its combination of Aristotelianism and anti-Aristotelianism, naturalism, and "secrets" traditions. The *De Magnete* is concerned with many of the same philosophical and epistemological issues as the other works I have been discussing and uses some of the same tropes and language. However, Gilbert's text is also the site of a significant epistemological shift. Unlike previous writers, Gilbert is not disturbed that the Copernican universe is at odds with the movements of the heavenly bodies as observed from the perspective of the Earth. Like Bostocke's argument that distillation provides evidence that is more "visible and palpable" than does ordinary experience of matter, Gilbert argues that the behavior of his "terrella" or magnetic model of the Earth makes the movement of the Earth "manifest," providing access to truth about its movement that is more accurate, and obvious, than ordinary experience of its apparent stillness. Gilbert is perhaps the first English natural philosopher to develop a practice of translation or mediation, as described by Bruno Latour, in which "the sciences do not speak of the world, but rather construct representations that seem to push it away, but also bring it closer."[82] Although Gilbert doesn't directly articulate the grounds on which he substitutes the experimental behavior of a model for direct observation of the universe itself, he carries out that substitution.[83]

Gilbert, like Digges, Dee, and Recorde, was seeking a way to make sense of the Copernican model of the universe, although in his case he was not focused on the structure and movement of the heavens but on terrestrial physics. As Gad Freudenthal has argued, Gilbert was trying to come up with a "suitable non-Aristotelian physics" to "complement the new, non-Ptolemaic cosmology."[84] We can recall that Aristotelian physics was based on the idea that each of the four elements had its own natural motion and *telos*: the element earth naturally sought the center of the universe, which explained the location of the Earth itself in the Ptolemaic universe, stationary at the center. If the Earth was not fixed at the center of the universe, then Aristotelian physics was wrong. There needed to be some other explanation for why the Earth "has a fixed orientation in space" and why it rotates on its axis.[85] Because magnets exhibited "rotational movements," Gilbert evidently believed that they might form the basis of an alternative theory of matter and motion.[86]

Like the other late-sixteenth-century writers that I have been tracing here, Gilbert hoped to resolve an epistemological gap that had opened up in the Aristotelian universe and to do so by focusing on an occult phenomenon: in this case, the archetypal example of an occult force, the magnet. Since the manifest qualities of the element earth—its heaviness, coldness, and density—no longer explained the location and movement of the planet Earth according to Copernicus, Gilbert turned to examination of an occult quality—magnetism—that seemed to inhere within the Earth itself. Gilbert revealed the sleight of hand that had to happen in order for empiricism to emerge from Aristotelian experience. He had to explain away the ordinary evidence of the senses (that earth is inert, that the heavenly bodies move and the Earth is still) and replace it with experimental evidence for counterintuitive truths (that the Earth is animated by the invisible force of magnetism, that it rotates on its axis every twenty-four hours, creating the illusion that the heavenly bodies move). Gilbert's work calls on the daily evidence of the senses when it supports his argument, and ignores or explains it away when it doesn't. In many cases he substitutes the visible evidence provided by the terrella, his spherical magnet, for what is actually observed on Earth. He often argues that it is more "plausible" or "reasonable" to accept that the senses are mistaken and even insists that the revolution of the Earth is "manifest."

His book signals early on its affiliation with the secrets tradition: "in the discovery of secret things and in the investigation of hidden causes, stronger reasons are obtained from sure experiments and demonstrated arguments than from probable conjectures and the opinions of philosophical speculators of the common sort."[87] Like the Paracelsan Bostocke, Gilbert assumes that true knowledge about nature comes from hidden, rather than manifest, phenomena. Unlike Bostocke, however, Gilbert emphasizes that he will begin with observation of the lodestone's manifest qualities: "the common stony and ferruginous matter, and magnetic bodies, and the parts of the earth that we may handle and may perceive with the senses" (1). Only then will he "proceed with plain magnetic experiments, and to penetrate to the inner parts of the earth" (1).

Like Digges and Harriot, Gilbert imagines that his new experimental techniques will allow him to visit places previously unavailable to human observers. He can "penetrate to the inner parts of the earth" and has "examined very many matters taken out of lofty mountains, or the depths of seas, or deepest caverns, or hidden mines." Just as geometry moves from "readily understood foundations" to "the highest and most difficult demonstrations whereby the ingenious mind ascends above the aether," so "magnetic doctrine" can reveal "things most secret and privy in the earth" (1).

On the one hand, Gilbert sees his work as related to an esoteric and secret tradition, commenting that "philosophy is for the few" (2). On the other hand, he anticipates seventeenth-century thinkers when he insists that other people should carry out his experiments in order to verify them, provided that they "handle the bodies carefully, skillfully, and deftly, not heedlessly and bunglingly." He also explicitly differentiates his work from the purposefully obscure writings of some purveyors of secrets, noting that "we sometimes employ words new and unheard of, not (as alchemists are wont to do) in order to veil things with a pedantic terminology and to make them dark and obscure, but in order that hidden things with no name and up to this time unnoticed may be plainly and fully published" (2). Gilbert presents himself as a mediator between previously hidden secrets and a wider public.

Scholars have tended to see Gilbert as most closely allied with the new science in his explicit debunking of the authoritative books of the past and of specific lore about magnets found in those books.[88] However, we can see that while Gilbert is more explicit and emphatic in his attitudes toward ancient authority than the writers who came before him, these attitudes have been anticipated by some of the writers we have been examining in these chapters. Like Gabriel Harvey, Gilbert defines as "true philosophers" those who "not only in books, but in things themselves look for knowledge" (1). Just as Recorde praised Ptolemaic astronomy while urging his pupil not to accept all of its tenets simply on authority, so Gilbert argues that to "the first parents of philosophy, to Aristotle, Theophrastus, Ptolemy, Hippocrates, Galen, be due honour rendered forever." He reminds us, however, that "our age has discovered and brought to light very many things which they too, were they among the living, would cheerfully adopt" (2).

In this vein, Gilbert is known for debunking in no uncertain terms "facts" inherited from writers like Pliny. About magnets, "for example, they asserted that a lodestone rubbed with garlic does not attract iron; nor when it is in the presence of a diamond" (3). He calls these "errors" that have spread like "evil and noxious plants." Although Gilbert is more emphatic than previous writers that such ideas are erroneous and dangerous, we can still recall Leonard Digges's tentative questioning of some meteorological facts derived from Pliny, as well as Fulkes's bolder attempt to find naturalistic explanations for as many of them as possible. Gilbert's invective against Aristotelian concepts resembles that of anti-Aristotelian polemicists writing in the Ramist or Paracelsan traditions; however, Gilbert differs from someone like Gabriel Harvey in actually dispensing with many basic tenets of Aristotelian naturalism. Gilbert has clearly turned a corner toward modernity in his relation to authority and experimentation, but we can also see that other sixteenth-century natural philosophers accompanied him right up to that corner.

Gilbert begins by arguing—as Bostocke and other Parcelsans suggested, and as

Aristotle's *Meteorology* also acknowledged—that the Aristotelian elements do not exist on Earth, at least not in their pure form: "the Aristotelian element, earth, nowhere is seen, and the Peripatetics are misled by their vain dreams about elements" (24). He argues that the surface of the Earth presents us with forms of earth—"loamy, or argillaceous, or sandy"—that have been changed and adulterated by weather and the heavenly bodies (as Aristotle would agree). The "primary and homogeneous earth-substance" is found only deep within the earth where we cannot access it. His main objection to the Aristotelian element earth is that it is "formless, inert, cold, dry, simply matter, the substratum of all things having no activity" (25). Contrary to Aristotelian dictum, Gilbert argues that this inert earth is not manifest, having "never appeared to any one, even in dreams, and if it did appear it would be of no effect in nature" (25). Like Bostocke again, Gilbert objects to Aristotelian elements because they are inert, unlike the "magnetic homogenic nature" of the interior of the Earth, which has an "energic form" (25, 24). For Gilbert, the observed movements of his experimental *terrella* can be "seen," while the inert element supposedly located inside the Earth cannot be.

Gilbert's main objection to Aristotle's theory of the elements is that he "admits only two simple movements of his elements—from the centre and toward the centre" (26). Instead, Gilbert argues that there are five natural movements exhibited by lodestones and therefore by the Earth itself: Coition, Verticity, Variation, Declination, and Revolution (26). Here again, Gilbert dismisses the ordinary observed experience of matter (that earthy things do fall and fire does rise) and accepts as more valid the motions exhibited by magnets (and the *terrellae*) in the course of various experiments.

Gilbert seems to be able to make this substitution (accepting the experimental motions of magnets as providing truer evidence about terrestrial physics than the ordinary observed behavior of earthly matter) through synecdoche. He believes that the behavior of a part of the Earth (the lodestone) provides valid evidence of the nature of the whole: "that the earth is fitted for circular movement is proved by its parts, which, when separated from the whole, do not simply travel in a right line, as the Peripatetics taught, but rotate also," as "a spherical lodestone, when floated in water, moves circularly on its center to become (as it seems) conformed to the earth on the plane of the equator" (112).[89]

Peter Dear has argued that, "for Gilbert, the *terrella* is the same as the earth in every way that matters."[90] I would go further and argue that Gilbert is here in the position of asserting that observing the behavior of the terrella, his spherical lodestone, tells us more about the behavior of the Earth than observing the Earth itself. The relationship between the terrella and the Earth is harder for him to articulate and seems to be analogical and based in traditional ideas about the perfection of the sphere.[91] He explains that he has created these spherical magnets because: "Inasmuch as the spherical form, which, too, is the most perfect, agrees best with the earth, which is a globe, and also is the form best suited for experimental uses, therefore we propose to give our principal demonstrations with the aid of a globe-shaped lodestone, as being the best and most fitting" (9). The spherical shape "agrees best with the earth," is "best suited for experimental uses," and therefore is "best and most fitting." Gilbert performs experiments designed to demonstrate the existence of the five natural motions of the Earth that he has posited: "all experiments that are made on the terrella, to show how magnetic bodies conform to it may—at least the principal and most striking of them—be shown on the body of the earth" (106). Interestingly, Gilbert hedges here on how exact is the correspondence between terrella and Earth.

After describing the experiments that demonstrate the five motions inherent in the Earth, Gilbert argues that the motions demonstrated by the terrella prove the Copernican hypothesis that the Earth revolves on its axis and that the observed movements of the heavenly bodies can actually be explained by the movement of the Earth. Gilbert's cosmology relies on ordinary observation when it suits him. He debunks the existence of the ninth and tenth spheres, as did Recorde, because they are not visible. He goes beyond Recorde to argue that no one has ever "found in one same (Ptolemaic) sphere those stars which we call fixed, or ever has given rational proof that there are any such adamantine spheres at all" (108). Similarly, he argues that the *primum mobile* "presents no visible body, it is in no wise recognizable" (108) and therefore is a "fiction" that is "purely a product of imagination and mathematical hypothesis" (109).

In a sharp (though only implicit) departure from Aristotelian epistemology, Gilbert implies that the Ptolemaic model of the universe is wrong because it was "adapted to the capacities of the vulgar," linking ordinary observed experience to a pejorative sense of social class. Gilbert argues that various heliocentric theories in antiquity were replaced by a geocentric theory "when philosophy had come to be handled by many, and had been given out to the public, then theories adapted to the capacities of the vulgar herd or supported with sophistical subtleties found entrance into the minds of many, and, like a torrent, swept all before them" (107). He thus condemns the Ptolemaic universe for being simultaneously too vulgar and too subtle.

He acknowledges that we observe the heavenly bodies to move but presents the movement of the Earth as a more reasonable explanation for this phenomenon than that our observations are accurate and the heavenly bodies do move: "But sure we see the sun, the moon, and the other planets and the whole heavenly host, within the term of one day come and depart, then either the earth whirls around in daily motion from west to east, or the whole heavens and all the rest of the universe of things necessarily speeds about from east to west" (108). Gilbert repeats an argument made by Copernicus and repeated by Thomas Digges, that because "nature ever acts with fewer rather than with many means," it is "more accordant to reason that the one small body, the earth, should make a daily revolution than that the whole universe should be whirled around it" (110). Therefore, Gilbert registers no concern about the discrepancy between perception and reality when "so are we whirled round with the earth while we think we stand still and are at rest" (109). While Digges felt the need to accompany this realization with the comment that "so many ways is the sense of mortall men abused" (*Prognostication*, Mr), Gilbert more readily accepts that reality is contrary to the evidence provided by the senses.

Like Digges, Gilbert is also happy to find an explanation for precession of the equinoxes that preserves the fixity of the stars, and here again he seems unbothered that the structure of the universe is other than it appears: "surely it must seem more probable that the appearances of the heavens should be produced by a deflection and inclination of the small body, the earth, than by a whirling of the whole system of the universe" (117). He is also unbothered that he cannot in the same way account for the movement of trepidation, noting only that "we cannot assign with certainty any natural causes for the motion" (121).

Gilbert essentially redefines what is to be considered "manifest," not only rendering the power of the magnet, which had previously been considered as the prime example of an occult force, to be manifest through experimentation, but also declaring that the invisible magnetic force that causes the Earth to rotate on its axis is "manifest" and "conspicuous." Having determined through experimentation that "a spherical lodestone, when floated in water, moves circularly on its center to become (as it seems) conformed to the earth on the plane of the equator," he concludes that the Earth "by an energy that is innate, manifest, conspicuous, revolves in a circle toward the sun; through this motion it shares in the solar energies and influences" (112). Gilbert here seems to perform a double sleight of hand; his terrella allows him to make the power of the lodestone visible through experiment and then to extend that power to the similarly shaped Earth and to declare it manifest.

His argument that the rotation of the Earth is a more "rational" and "probable" alternative than the rotation of the heavens is shored up by the animist tendencies of his magnetized Earth. He argues that if the Earth did not rotate, one side would be scorched by the sun while the other froze. Therefore "as the earth herself cannot endure so pitiable and so horrid a state of things on either side, with her astral magnetic mind she moves in a circle" (112).

Gilbert thus seems of a piece with other late-sixteenth-century English natural philosophers in his appeal to a combination of experimentation on occult phenomena and appeal to secrets traditions in order to find a way to accept the Copernican model of the universe. Even his polemic against the authority of Aristotle and Ptolemy has antecedents. What differs is the substitution of the observed behavior of an artifact for the ordinary observation of the physical world and the assertion that the behavior of the artifact can be mapped onto the visible world, rendering a phenomenon that cannot be directly observed—the rotation of the Earth—to be manifest. The effort that it takes for Gilbert to make this epistemological leap is apparent in his varying attempts to explain the relationship between the terrella and the Earth, and in his various appeals to reason, probability, and the psychology of an animate Earth. That he can discount without difficulty "what daily we see here at eye" marks a true paradigm shift and reveals the extent to which empiricism had to grow out of a paradoxical redefinition of what constitutes the "manifest."[92]

CHAPTER 4

Spenser and the New Science

Edmund Spenser has not generally been credited with knowledge of, or interest in, the changing ideas about the natural world that were developing during his lifetime.[1] Allison Chapman has linked his *Shepheardes Calender* with controversy over calendrical reform in the period, but she emphasizes the political rather than natural philosophical issues at stake.[2] His *The Faerie Queene* has long been associated with archaic modes of thought and writing, looking back to classical antiquity and the Middle Ages rather than forward to new ideas about the natural world.[3] He has generally been seen as engaging most fully with moral and political philosophy, interested largely in questions of ethics, law, sovereignty, and human relationships.[4] However, he was aware of a number of the natural philosophical and epistemological issues that arose as the Aristotelian synthesis fell apart, and *The Faerie Queene* in particular is deeply informed by his increasing concern about the intelligibility and stability of the universe at the end of the sixteenth century. While Spenser's works reflect the confusion and anxiety of his age, ultimately *The Faerie Queene* raises the threat of new ideas only to fall back on Aristotelian orthodoxy.

Spenser's reputation for ignorance about natural philosophy stems in part from comments made by his friend (or rather frenemy) Gabriel Harvey. In his marginal comment on an English translation of a (probably third-century) Greek cosmographical poem, Dionysius Perigetes' *The Surveye of the World* (translation by Thomas Twinne, published in 1572), Harvey praises English poets like Geoffrey Chaucer and John Lydgate, whose works include "Notable Astronomical descriptions . . . much better learned than owre moderne poets"; Harvey mentions especially "Chawcers conclusions of the Astrolabie."[5] He praises a number of passages in Chaucer, including descriptions of weather, seasons, times of day, astrologers, and magicians. He goes on to say, "It is not sufficient for poets, to be superficial humanists: but they must be exquisite artists, and curious universal schollers" (161). He comments then that "M. Digges hath the whole Aquarius of Palingenius bie harte" (161), perhaps not realizing the gap between Digges's own astronomical theory and that found in Palingenius's 1531 *zodiacus vitae*, a Latin poem that uses the signs of the zodiac as an organizing structure to survey all learning.

In contrast, Harvey points out Spenser's ignorance (the original is in Latin): It is a shame that "Spenser himself, even if he isn't completely ignorant of the Sphere and the astrolabe, is unlearned in astronomical laws, tables, and instruments. Especially, since he has seen our Blagrave's Margaritam Mathematicam. Which yields neither to Pontanus, or Palingenius, or Buchanan, or even Bartasius, in the excellent knowledge of his Globe, astrolabe, and familiar staff. So that he seems to be another Digges, or Harriot, or even Dee."[6] Harvey is here mostly advocating for didactic poetry of the kind written by Palingenius and wishing that Spenser would write scientific poetry. His list, which seems to equate Blagrave (who wrote practical treatises on the use of astronomical instruments in navigation) and Palingenius (author of an encyclopedic Latin poem containing one section on astronomy) with Dee and Harriot, who were on the forefront of speculative thought, suggests that Harvey himself lacks a clear understanding of the purpose of these works and is not really in a position to judge Spenser's level of knowledge.[7]

Although Spenser for the most part shunned Harvey's advice and failed to write a didactic poem on mathematical instruments, *The Faerie Queene* was strongly influenced by the breakdown of Aristotelian naturalism and the epistemological issues that accompanied it. Although Spenser most explicitly engages with scientific issues later in the poem, in Book V of *The Faerie Queene* and in the *Mutabilitie Cantos*, his allegorical method throughout the poem was, I believe, partly influenced by the uncertainty about the legibility of nature expressed by the English writers of natural philosophical treatises during the sixteenth century. Spenser twice takes on current discussions about the relationship between the supra- and sublunar realms, the fixity of the stars, and the irregular motions of the planets. In Book V, he seems to interpret the phenomenon of precession as a sign that the cosmos is in decline and the order of the elements on earth under threat; he does this probably in the context of apocalyptic reactions to the new star of 1572.[8] Here he tries out the idea that scientific experimentation with a balance as outlined by John Dee in the preface to Euclid could restore order to the terrestrial elements, but he ultimately has Artegall resoundingly reject this solution in favor of a reiteration of orthodox Aristotelianism.

In the *Mutabilitie Cantos*, Mutabilitie raises more forcefully the possibility that there is no distinction between the areas above and below the moon, arguing that phenomena such as precession and planetary retrogradation prove that change is possible in the supralunar realm. Here, Spenser uses two tropes familiar from the treatises that I discussed in chapter 3. First, Mutabilitie travels up to the Moon and then to the higher heavens. This trip to a previously inaccessible realm reveals no new information that can resolve the problems that Mutabilitie raises, so Spenser resorts to an earlier device: Nature herself speaks in order to provide instruction about the questions raised. Like Artegall, Nature denies the evidence of phenomena that trouble the Aristotelian universe and uses her authority alone to assert that orthodox Aristotelian theories are correct. Spenser here also explicitly takes on questions of whether Nature presents its truths in a manifest or occult manner, ultimately choosing a paradoxical and poetic conclusion: she is "unseen of any, yet of all beheld."[9]

A concern with manifest and occult signs informs Spenser's allegory throughout the poem and functions as one of its most powerful engines. Because Temperance is a medical as well as ethical concept, I look at Book II of *The Faerie Queene* in order to trace the ways in which characteristic Spenserian scenes of interpretation and failed interpretation emerge in part from the contemporary concerns about whether the observed surface of nature (and of the human body) provide manifest access to truth. Many scholars have, of course, charted the characteristic Spenserian mode of allegory, which questions the transparency of its own signification. And there are certainly many intellectual contexts that might contribute to our understanding of Spenser's interest in the difficulties of interpretation. I am interested in showing that the breakdown of an intuitive relation to the natural world constitutes one strand of Spenser's obsession with reading the surfaces of the world.

Lauren Silberman has argued that in Book I, the hero "is educated in not trusting his senses" because "trouthe—faith and loyalty—is shown to be superior to truth, conceived as empiricism." Empiricism fails in Book I because "the fallen world is properly seen as a series of signs pointing to the higher truth available to faith alone."[10] She argues that Book II "takes up the project of coping with, rather than transcending, the fallen world," and I would argue that in the context of this fallen world, contemporary concerns about the legibility of the universe on its own terms replace (or intensify) concerns about the legibility of God's plan in the universe.[11]

SPENSER AND ASTRONOMY

We have seen how equinoctial precession and planetary retrogradation posed problems for Ptolemaic astronomy, since the slow movement of the fixed stars through the signs of the zodiac and the apparent backward movement of the planets were difficult to account for under that system. The "New Star" of 1572 lent urgency to the problem of precession for some people because it suggested that change could, in fact, occur in the realm of the fixed stars. Copernicus solved precession and retrogradation by explaining what appeared to be anomalous heavenly motions as actually functions of the motion of the Earth. However, he did so at the cost of asserting that the stillness of the Earth and the observed movements of the heavens were illusory. As we have seen, educated English writers in the last twenty-five years of the sixteenth century took a range of positions on these fraught issues. It should not be a surprise, then, that Spenser also felt the need to think through the ways in which the world of *The Faerie Queene* was impacted by these contemporary problems in natural philosophy.

Spenser's proem to Book V offers a pessimistic account of the decline of the world, calling on narratives of its successive "ages," from an initial golden age to a current degenerate "stonie one," found in various classical sources. However, he introduces a new twist when he links the conventional Ovidian account of social decline to disorder in the heavens, taking four stanzas to describe the precession of the equinoxes:

> For who so list into the Heavens looke,
> And search the courses of the rowling spheares,
> Shall find that from the point, where they first tooke
> Their setting forth, in these few thousand yeares
> They all are wandered much.[12] (V.pr.5.1–5)

Spenser goes on to refer to the obliquity of the ecliptic in some detail: the sun

> is miscarried with the other Spheres.
> For since the terme of fourteene hundred yeres,
> That learned *Ptolomae* his hight did take,
> He is declined from that marke of theirs,
> Nigh thirtie minutes to the Southerne lake.[13]

Spenser also refers to the irregularity of planetary orbits, noting that "most is *Mars* amisse of all the rest, / And next to him old *Saturne*, that was wont be best" (V.pr.8.8–9).

Astronomical observers had been aware of precession and the irregularity of planetary orbits since antiquity. However, as we have seen in treatises like Recorde's *Castle of Knowledge*, these phenomena had been "explained" through elaborations of Aristotle's original cosmology. Ninth and tenth celestial spheres were added to the system to account for the motion of the fixed stars, and a complex system of epicycles was added to account for the irregular motion of the planets. According to orthodox Aristotelian-Ptolemaic theory, there would be no reason to associate precession with "wandering" or decline. However, the pressing issue of calendar reform in the sixteenth century led Copernicus, among others, to reconsider the phenomenon and to offer an alternate explanation for it. Spenser, in this passage, shows off a technical knowledge of precession, probably gained through his involvement in the calendar debate.[14] For Spenser, writing after the appearance of a new star in 1572 and evidently unaware of or unwilling to accept the Copernican explanation, precession becomes a sign of the beginning of the end of the Aristotelian universe and therefore marks a threat to its stability and intelligibility. Spenser was probably moved to take a more pessimistic view of the phenomenon of precession as a sign of instability, unintelligibility, and decline because he was aware of observations made "of late" that proved the realm of fixed stars was subject to change.[15]

Into this fallen universe, Spenser introduces Artegall, whose task it will be to restore justice to it. Canto I explains how Artegall "in justice was upbrought / Even from the cradle of his infancie, / And all the depth of rightfull doome was taught" by Astraea herself (V.i.5.1–3). Before leaving the Earth in disgust at man's wickedness, she teaches Artegall "to weigh both right and wrong / In equall balance with due recompence / And equitie" (V.i.7.1–3). However, when Artegall later (in Canto ii) encounters a giant who is attempting to use a balance to restore the original proportions of the elements, he insists that the Earth is not in decline and that right and wrong cannot be weighed in a balance. The Giant is then pushed off a cliff to his death.

THE SCIENCE OF WEIGHT

The Giant's attempt to use a balance to restore the proportion of elements in the world is directly related, in several ways, to the developments in science to which the proem alludes. Like William Gilbert, Spenser seems to have realized that an altered theory of the heavens implied alterations in terrestrial physics. We should recall that the position of the Earth at the center of the Ptolemaic universe is a function of its weight: according to Aristotle, the Earth must be at the center because the heavy element of earth naturally seeks the center. In his rebuttal of the Giant, Artegall provides a clear account of Aristotelian Orthodoxy.

> The earth was in the middle centre pight,
> In which it doth immoveable abide,
> Hemd in with waters like a wall in sight;
> And they with aire, that not a drop can slide:
> Al which the heavens containe, and in their courses guide. (V.ii.35. 5–9).

Under ideal circumstances, the four elements would arrange themselves as Artegall describes, in concentric spheres of earth, water, air, and fire. However, according to Aristotle, weather on Earth, caused by the influence of the sun and planetary spheres, disturbs this ideal arrangement to create the conditions that we experience: wind and water disturb and redistribute earth. Ultimately, these disturbances are kept under control by the influence of the fixed stars and by the teleology of the elements themselves ("al which the heavens containe, and in their courses guide").

If, however, the structure of the universe is questioned, the stability of the elements might also be open to question. If earth doesn't naturally seek the center, what is to keep it from flying off into the heavens? If the fixed stars are not fixed, what lends stability to the system? If, as Spenser acknowledges in the proem, "the heavens revolution / Is wandred farre from where it first was pight" (V.pr.4.6–7), how can Artegall maintain that the Earth does still "immoveable abide" where it was "pight"? The Giant's belief that the balance of elements on Earth has become disturbed would be a plausible corollary of the astronomical disturbances described in the proem. The Giant's use of a balance to attempt to restore the order of the elements on Earth might seem like a similarly plausible use of a scientific instrument to understand and master the material world.

In his preface to Euclid, Dee includes an account of "Statike," "an Arte Mathematicall, which demonstrateth the causes of heavynes, and lightnes of all thynges: and of motions and properties, to heavynes and lightnes, belonging."[16] Dee lists "Statike" among branches of mathematics that are derived "from the puritie, absolutnes, and Immaterialitie of Principall *Geometrie*" and applied to measurement of material things (A3v). Dee is concerned to relate most human endeavors to mathematics, and the description of the science of weight is preceded by a claim (based on passages from Aristotle's *Nichomachean Ethics*) that Justice is dependent on arithmetic: "Nor (Justice in infinite Cases) without due proportion (narrowly considered) is hable to be executed," especially by someone who must "decide, with aequitie, the infinite varietie of Cases" (A1r, v). Dee also provides a history of practical geometry, arguing that it arose in antiquity to re-measure boundaries "when Boundes, and meres of land and ground were lost" whenever some force "did wrongfully limite, measure, encroach, or challenge (by pretence of just content, and measure) those landes and groundes" (A2r–v). Spenser's Giant similarly seeks to redress elements that have "encroched uppon others share."[17]

Dee presents "Statike" as an experiment-based science: "by the Bilanx, or Balance (as the chief sensible Instrument,) Experience of these demonstrations may be had: we call this Art, *Statike*: that is, *the Experimentes of the Balance*" (B4r–v). Dee provides a summary of some of Archimedes's findings about buoyancy and then (most probably) alludes to experiments proposed by Nicholas of Cusa to determine the proportions of the elements: "Moreover, by the foresaid propositions (wisely used.) The Ayre, the water, the Earth, the Fire, may be nerely, knowen, how light or heavy they are (Naturally) in their assigned partes: or in the whole. And then, to thinges Elementall, turning your practise: you may deale for the proportion of the Elementes, in the thinges Compounded" (C1r). The Giant's project of determining the original proportions of the elements does not seem so fanciful in light of this claim. Dee seems to equate the mental "weighing" of evidence in the mind with the experimental truth provided by the balance: "Frende, I pray you, way these thinges, with the just Balance of Reason. And you will finde Mervailes, upon Mervailes: and esteem one Drop of Truth (yea in Naturall Philosophie) more worth, then whole Libraries of Opinions, undemonstrated: or not aunswering to Natures Law, and your experience" (C1r).

Dee then provides directions for using the balance to solve otherwise difficult or impossible geometric problems involving the comparative volumes of solids; fill each solid object with water and compare the weights of the amount of water that it takes to fill each one. Here again, Dee sees the balance as a means of revealing truth "which, long hath bene wished for" by bridging the ideal world of mathematics with the material world of matter and mass: "And though it be Naturally done and Mechanically: yet hath it a good Demonstration Mathematicall" (C3r).

Despite these optimistic claims about the access to hidden truth provided by the balance, Dee is moved to interrupt his optimistic account of experimental science with a religious interjection that sounds like Artegall and echoes some of the Bible verses to which he refers. Dee himself apologizes for this digression, which he calls a "pang": "Mervaile nothing at this pang (godly frend, you Gentle and zelous Student.) An other day, perchaunce, you will perceive, what occasion moved me"(B4v). After urging readers to be aware of "what proffit, (all maner of wayes) by this Arte might grow," Dee addresses God, "Thou onely, knowest all thinges precisely (O God) who hast made weight and Balance, thy Judgement: who hast created all thinges in *Number, Waight, and Measure*: and hast wayed the mountaines and hils in a Balance: who hast peysed in they hand, both Heaven and earth" (B4v). "Peised" is another word that appears in Spenser's account of the Giant and nowhere else in his poem.

For Dee, the balance provides a way to gain access to aspects of nature that are hidden from view and to quantify complex phenomena. Without explicitly doubting the Aristotelian system, he nevertheless looks forward to modern experimental science by suggesting that the ordinary evidence of the senses needs to be supplemented with data provided by a scientific instrument if the "hidden" truths of nature are to be discerned.[18] The balance also provides a means of comparing very different kinds of things to each other, potentially in violation of Aristotelian mathematical essentialism. The need to discern these hidden causes becomes even more pressing after Aristotle's essentialist worldview began to be questioned. If the observed nature of things doesn't reveal their essence, how do you connect appearance with reality? One way to make this connection, as William Gilbert's work will reveal, is through experiment and analogy. If the causes and real nature of things are no longer what they seem, you must learn to understand them by analogy with what you can see. Thus, although we cannot observe how the hidden magnetic core of the Earth causes it to revolve on its axis, we can understand this motion by analogy with the circular movement of the terrella. In the early modern period, a corollary of the general abstraction from essential qualities was an expansion of what was considered to be commensurable; things are no longer identical to themselves, but they can be understood by comparison with very different things.[19]

Whether or not Spenser knew Dee's *Preface* (and I believe that verbal echoes suggest he did), the Giant's use of balances to redistribute the elements offers enough technical detail about the process to make us view it in relation to the science of weight. Thus, although the biblical analogues that Spenser brings to bear call his enterprise into question, its scientific provenance suggests that it could be imagined as an appropriate response to a disordered universe. If scientific discovery has rendered the universe unstable and unintelligible, one possible way to restore intelligibility and stability would be through scientific study. Spenser was attracted to this possibility and terrified by it.

ALLEGORY AND EXPERIMENT

Throughout *The Faerie Queene*, Spenser presents us with a world—landscape and characters—which, he repeatedly tells us, ought to be intelligible but rarely is. He uses allegory as a way of rendering the surface of his imagined world intelligible by force: A character is named Una and represents the singleness of truth. However, as most readers have noticed, the intelligible surface of Spenserian allegory refuses to stay that way: Una is doubled by a false spirit Una, and then again by Duessa. Redcrosse Knight cannot tell the real Una from her constructed double, and virtually no one can tell the true Florimell from the false one. The poem repeatedly tells its characters (and readers) that they should be able to "read aright," but allegory repeatedly breaks down into more complex and confusing figures like metaphor. As Annabel Patterson has noted, the balance is introduced first as a simple, stable, and "highly abstract" icon of justice but becomes more complex when, in the hands of the giant, it becomes an instrument of more material, and less exact, comparison.[20] The Giant-scientist is also a poet, who, like Spenser, uses metaphor in an attempt to capture the complex truth of the material world. Neither the Giant nor Spenser is able to reconcile an idealized sense of stability and intelligibility with actual experience of the material world.

When Spenser introduces Artegall's lessons in using the balance in Book V, it takes its familiar form as an abstract symbol of justice. Scales or balances were a common symbol of justice from antiquity, and as such they provide Spenser with a stable referent for his allegory.[21] The analogy between weighing in a balance and dispensing justice seems unambiguous. However, in this image, as in the traditional image of the scales of justice, what is being weighed is not clear. Spenser says "to weigh both right and wrong/in equall balance" (V.i.7.1–2), so it isn't clear whether "right" is in one pan of the balance and "wrong" in the other or whether "both right and wrong" are weighed separately—two rival "rights" against each other, or two rival "wrongs." In general, what is salient in the image of the balance is not what is being compared but the fairness of the mechanism as an instrument of comparison.[22] In traditional iconography, the scales of justice are often depicted as empty; they offer the potential to compare two things, but what they compare is not important. A related tradition depicts the weighing of souls at the last judgment, where a soul is weighed against collected sins, good deeds, or the gospel—again, what is compared is of less importance than the act of unbiased comparison.[23] As an allegorical image of justice, the balance in Canto i represents the act of comparison as unproblematic.

When the image of the balance returns later in Book V, Canto ii, Spenser shifts our attention to what is being weighed and calls attention to the problematic nature of comparison. This image functions less as an allegorical emblem and more as a complex analogy for the use of an artificial instrument to determine truths about the natural world and for the use of analogy itself to compare unlike things to each other. In Canto ii, Artegall encounters a giant who is "holding forth on hie / An huge great paire of ballance in his hand, / With which he boasted in his surquedrie, / That all the world he would weigh equallie" (V.ii.30.2–4). Like Dee's investigator of "Statike," the giant will use his balances to compare the proportions of the elements, variously offering to weigh sea against earth, fire against air, heaven against hell, and thereby repair the damage that has occurred over the years as each "had encroched upon others share" and made their proportions unequal. Like Dee, the Giant believes that weighing can give insight into human behavior ("so were realmes and nations run awry" [V.ii.32.6]) and potentially rectify it.

Artegall echoes Dee's brief religious "pang" when he strenuously objects to this project, arguing that the Giant's attempt to "weigh the world anew, / And all things to an equall to restore" is "great wrong" (V.ii.34.1–2). Artegall argues that the elements have been so bounded by God that "mongst them al no change hath yet beene found" (V.ii.36.4), directly contradicting the proem's claim that the universe has altered and declined. Like Dee, Artegall points out that the elements were all created "in goodly measure, by their Makers might, And weighed out in ballaunces so nere, That not a dram was missing of their right" (V.ii.35.2–4). As Judith Anderson notes, the Giant counters with an appeal to the evidence of the senses. Like an orthodox Aristotelian, he directs Artegall to "plainely see" (V.ii.37.4) how the sea does "encroch" upon the earth. The Giant accepts the Aristotelian faith in appearances, but they lead him to question the stability of the universe. Like Thomas Digges and Dee, the Giant believes that scientific discovery can create an alternative and more compelling order in the universe; unlike Digges, he does not believe that the senses are deceptive, arguing, instead, like Dee, that an instrument like the balance can reveal hidden truth and restore order.

Artegall, on the other hand, prefers stability to intelligibility. He contradicts the evidence of the senses to insist on an idealized version of order. Anderson rightly argues that Artegall's preference for mind over matter is Neoplatonic, and it resembles a passage in Dee's preface where he compares astronomy and music as "sister" arts of the eye and ear, respectively, but as arts in which the evidence of the senses must be perfected by the mind: "as *Astronomie* hath a more divine contemplation, and commodity, then mortall eye can perceive; so, is *Musike* to be considered, that the Minde may be preferred over the eare" (b2v).

But Artegall also insists on an orthodox Aristotelian essentialism. It is impossible, he argues, to compare unlike things such as truth and falsehood or right and wrong. Artegall also refuses to accept that an artificial instrument can provide access to truth about the material world. The balance cannot actually weigh right and wrong; it is just a symbolic representation of an act of judgment that must be carried out in the human mind. The balances "doe nought but right or wrong betoken" (V.ii.47.5). Unlike writers such as Robert Recorde and Leonard Digges, Artegall inhabits a post-Aristotelian world in which the "tokens" presented by the face of nature cannot provide reassurance of its stability and intelligibility. He finds himself in a worst of all possible epistemological worlds. Although, like Dee and Thomas Digges, he uses Neoplatonic idealism to counter his anxiety about the complex unreadability of nature, he isn't able to share their optimism that new mathematically informed approaches to nature can—as the Giant argues—restore its stability. The violent destruction of the Giant (whose own weight, and the laws of force and momentum cause his downfall) provides the only possible outlet for Artegall's epistemological impasse.[24]

The Giant's balance is thus a scientific instrument, but it is also a figure for a new kind of analogy, like that constructed by Gilbert between his terrella and the Earth, which needed to be developed as the universe no longer yielded up its truths directly. The gap opened up by the end of the Aristotelian era was precisely a gap between the immaterial and invisible causes of things and their material manifestations, between a theory of nature and the piecemeal material manifestations of it yielded up by experiment. This was a gap that could only be bridged by analogy, and, as Lorraine Daston and Dedre Gentner have shown, thinkers as different as Galileo and Boyle used analogy in careful, targeted ways to link theory with practice.[25] What was lost—a sense of being in direct contact with the truth about a stable natural world—was compensated for by an expansion in what could be productively connected by analogy or comparison.

MUTABILITIE AND CHANGE ABOVE THE MOON

Spenser must not have been fully satisfied with Artegall's violent refutation of the problems raised by the breakdown of the Aristotelian universe because he takes up the issues of precession and planetary orbits again in the *Mutabilitie Cantos*. Here, Spenser comes close to doing as Gabriel Harvey advised, writing a scientific poem. The *Cantos* are essentially about the contemporary debate over the Aristotelian division of the cosmos into a sublunar realm subject to change and a supralunar realm considered to be changeless. We can recall that Copernicus continued to accept this division of the cosmos into two fundamentally different realms but that it was eroded by the appearance of the new star in 1572 and discarded by William Gilbert in 1600.[26] However, Spenser treats these issues in his own way, translating astronomy into mythological narrative. Robert Schuler has argued that Philip Sidney's *Apologie for Poetrie* rejected didactic poetry in favor of "mimetic, morally significant poetry," and Spenser seems to be siding with Sidney rather than Harvey on this issue.[27] Rather than providing a descriptive account of the controversy (as Harvey seemed to advocate), Spenser's poem creates a fictional narrative about a figure called Mutabilitie who tries to convince the classical gods and Nature that she should have sovereignty above as well as below the moon. Here, Mutabilitie travels up into the heavens, not in order to learn hidden truths about the cosmos but in order to stake her claim to the supralunar realm. Nature appears not to provide information about previously hidden truths but, like Artegall, to assert the stability of the universe in the face of clear evidence to the contrary. In the *Cantos*, Spenser relies on paradox to resolve questions about whether nature's truths are manifest or occult; in his poetic formulation they can be both at once, although true knowledge is ultimately deferred until the end of time.

The argument of the *Two Cantos of Mutabilitie* explicitly identifies its topic as the issue of whether change can occur above as well as below the moon: "Proud Change (not pleasd, in mortall things, / beneath the Moone to raigne) / Pretends, as well of Gods, as Men, / to be the Soveraine." Mutabilitie is then introduced as a Titaness who is the cause of change and mortality on earth. As Spenser in Book V of *The Faerie Queene* had tied the alteration represented by precession to classical myths about the loss of a golden age, he links Mutabilitie's sovereignty on earth with the biblical fall:

> For, she the face of earthly things so changed,
> That all which Nature had establisht first
> In good estate, and in meet order ranged,
> She did pervert, and all their statutes burst:
> And all the worlds faire frame, (which none yet durst
> Of Gods or men to alter or misguide)
> She alter'd quite, and made them all accurst
> That God had blest. (VII.vi.5.1–8)

Traveling up through the concentric regions of "ayre" and then "fire," she makes her way to the "Circle of the Moone" (VII.vi.7, 8) and tries to shove Cynthia from her throne, causing an eclipse.[28] Unlike scientific writers who imagined that mathematical "wings or stairs" up to the heavens could provide evidence about the structure of the universe, Mutabilitie does not travel in search of knowledge. She believes she already knows as much as she needs to know about astronomy—that the physics of the supralunar realm is continuous with that of the sublunar realm such that change is possible throughout the universe. When Jove denies her claim, she makes a legal argument that only "the God of Nature" (VII.vi.35.6) can adjudicate her case.[29] Her quest is then interrupted by a digressive narrative about the location on Earth where Nature will consider her case. This story about Diana's abandonment of "Arlo-hill," or Galtymore near Kilcolman Castle in Ireland, echoes Astraea's abandonment of Earth in Book V. Spenser seems to associate disruptive changes in the heavens with loss, imagined as abandonment by a female figure who had previously hallowed and stabilized the Earth. It is as if the ordering force supposedly exercised upon the Earth by the fixed and orderly heavenly bodies under the Ptolemaic system is imagined as a female deity, whose ascension into the heavens leaves the Earth disordered and bereft. That Diana leaves the Earth to live on the moon, guarding the border between change and permanence, supports this interpretation. Mutabilty's ascension into the heavens (where she doesn't belong) compounds and extends the disorder originally caused by Diana's abandonment of the Earth.

Mutabilitie begins her argument to Nature with the idea that the gods are oppressing the creatures of the cosmos "with power, unequally," and A. C. Hamilton in a marginal note likens her to "the Giant with the Scales."[30] She first describes her power over the elements in the sublunar realm in terms consistent with Aristotle's *On Generation and Corruption* and *Meteorology*.[31] Mutabilitie asserts that the elements constantly change into one another, but, like many early modern natural philosophers, she views these transmutation phenomena as "wondrous slights," difficult to explain in Aristotelian terms.

> Thus all these fower (the which the ground-work bee
> Of all the world, and of all living wights)
> To thousand sorts of *Change* we subject see:
> Yet are they chang'd (by other wondrous slights)
> Into themselves, and lose their native mights;
> The Fire to Aire, and th'Ayre to Water sheere,
> And Water into Earth: yet Water fights
> With Fire, and Aire with Earth approaching neere:
> Yet all are in one body, and as one appeare. (VII.vii.25)

In the course of these claims about sovereignty over the earth, she briefly seems to acknowledge the Copernican argument that the Earth itself moves although it doesn't appear to: "And first, the Earth (great mother of us all,) / That only seems unmov'd and permanent, / and unto *Mutability* not thrall; / Yet is she chang'd in part, and eeke in general" (VII.vii.17.6–9). Hamilton glosses this: "Mutabilitie may be expected to support the new astronomy." I think it is possible that Mutabilitie here refers to the alterations of the elements on Earth that she goes on to describe and not to Copernican theory. However, the phrase "that only seems unmov'd" resembles Thomas Digges making the opposite point that "the Sunne and the great Globe of fixed starres seems to sway about and turne, albeit indeed they remain fixed. So many wayes is the sense of mortall men abused." If she does here refer casually to Copernican theory, it would be the only place where Spenser indicates that he understood its implication that the Earth is in motion.

Following lengthy descriptions of the change of seasons and months on Earth, Jove readily cedes Mutabilitie's control over the sublunar region but insists that it cannot extend into the heavens above the moon. Mutabilitie argues in response that she can "prove, that even yee / Your selves are likewise chang'd, and subject unto mee" (VII.vii.49.8–9). Reversing the order and emphasis of the proem to Book V, Mutabilitie describes the irregular motions of the planets at some length and then gives a brief account of precession. She emphasizes planetary motion because the planets share the names of classical gods, and her argument is literally that the gods "your selves are likewise chang'd." She describes how Mercury "his course doth alter every yeare, / And is of late far out of order gone" (VII.vii.51.3–4). Russell Meyer explains the comment "of late" by referring to the discrepancy between the old Alfonsine tables for locating the planets and stars and their replacement by the Prutenic tables based on Copernican calculations (which were still inaccurate).[32] As I noted in a previous chapter, Harvey seems to have seen Copernican theory as significant largely for its revision of the Alfonsine tables: "Changed judgements: since the fixed stars have changed places. The old tables of Ptolemy and of others are not congruent with today's observed phenomena: it is very clearly evident from the Copernican revolutions and the Prutenic tables of Reinhold, where the errors of the old tables ought to be reformed."[33] Spenser may well have shared this view that Copernican theory established that the heavenly bodies had changed position over time.

Mutabilitie continues to press her case, arguing that "*Mars,* that valiant man is changed most: / For he, some times so far runs out of square, / That he his way doth seem quite to have lost" (VII.vii.52.1–3). As a result, astronomers are "stonisht" and "damne their lying books" (VII.vii.52.5, 6). She at first seems to acknowledge the fixity of the stars—"onely the starrie skie doth still remaine"—but then argues that precession alters even that realm: "Yet do the Starres and Signes therein still move, / And even it selfe is mov'd, as wizards sain. / But all that moveth, doth mutation love" (VII.vii.55.5–6, 8). With this conclusive argument, she rests her case: her sphere of influence extends above the moon to impact the planets and even the stars.

Nature, like Artegall in Book V, renders judgment against Mutabilitie based on orthodox Aristotelian principles. Everything changes in order to move toward its ordained end.

> I well consider all that ye have sayd,
> And find that all things stedfastnes doe hate
> And changed be: yet being rightly wayd
> They are not changed from their first estate;
> But by their change their being doe dilate:
> And turning to themselves at length againe,
> Doe worke their owne perfection so by fate:
> Then over them Change doth not rule and raigne:
> But they raigne over change, and doe their states maintaine. (VII.vii.58)

Mutabilitie has presented evidence that the Aristotelian doctrine of changelessness above the moon has been found to be false, and she argues that this fact would have implications for the stability of the elements below the moon, potentially disrupting their teleological ordering. Nature simply disregards this evidence.[34]

Many stanzas have been devoted to Mutabilitie's presentation of evidence in support of her claim to sovereignty above the moon, and her preponderance of evidence surely outweighs the two stanzas that Nature devotes to dismissing her arguments. Nature does not attempt to refute this evidence but simply uses her authority to assert the Aristotelian theory that is already crumbling. Nature, like Artegall, wins the argument by assertion of superior authority, and as a result, Spenser's second try at resolving these issues is no more satisfying than the first. Several critics point out that the fragmentary "unperfite" eighth canto counters Nature's assertion that all things "doe worke their owne perfection so by fate," deferring her vision of stability to the end of time.[35]

THE INTELLIGIBILITY OF NATURE

The *Cantos of Mutabilitie* also confront the issue of whether Nature is intelligible and fail to provide an unambiguous resolution to this problem. Spenser's representation of Nature implicitly raises the question, current in late sixteenth-century scientific writing, of whether truths about the natural world are manifest or occult. When Nature first appears, we learn that her face is not visible. We can tell by her size that she is more powerful than any other being, but we cannot discern any other facts about her by her appearance:

> Then forth issewed (great goddesse) great dame *Nature*,
> With goodly port and gracious Majesty;
> Being far greater and more tall of stature
> Then any of the gods or Powers on hie:
> Yet certes by her face and physnomy,
> Whether she man or woman inly were,
> That could not any creature well descry:
> For, with a veile that wimpled every where,
> Her head and face was hid, that mote to none appeare. (VII.vii.5)

Spenser spends two stanzas speculating about why her face is veiled. Is it because she is too frightening, or too beautiful to be seen by mortals?[36] He then offers a curious excuse for not describing her in more detail: even Chaucer failed to describe Nature in his *Parliament of Foules*, pointing readers instead to Alan of Lille's *De Planctu Naturae*, a text that seems not to have been available to Spenser.[37] It seems possible that Spenser is getting in a dig at Gabriel Harvey, who praised Chaucer's descriptions of nature in contrast with Spenser's ignorance of it. If Harvey's marginalia reflects sentiments that he expressed to Spenser, urging him to write descriptive didactic poetry like Chaucer, Spenser may here be pointedly indicating Chaucer's failure to describe Nature herself.

Whether or not this passage alludes to Harvey's ideas about scientific poetry, it does reflect contemporary concerns about the ways in which the causes of natural phenomena seemed increasingly disconnected from the manifest appearance of nature. Thus, the Earth that "only seems unmov'd" (VII.vii.17.7) and the "wondrous slights" (VII.vii.25.4) that cause transmutations of the elements provide examples of the epistemological uncertainty represented by Nature's veiled face. When Mutabilitie responds to Jove's argument that change cannot occur in his realm above the moon, she argues that:

> The things
> Which we see not how they are mov'd and sway'd,
> Ye may attribute to your selves as Kings,
> And say they by your secret powre are made:
> But what we see not, who shall us perswade? (VII.vii.49.1–5)

Hamilton glosses this as an allusion to 2 Corinthians 4.18: "The things which are not sene are eternal."[38] But Mutabilitie surely also voices an insistence that she will accept only manifest evidence and will not consider occult causes ("secret powre") as valid. Here she sounds like Robert Recorde's Master in the *Castle of Knowledge*, who will not accept the theoretical ninth and tenth spheres posited as an explanation of precession because they are not visible: "I thinke it best to tell you of no mo spheres, then are perceptible by sighte, for so manye are we certaine of" (7).

When Nature delivers her final judgment that, contrary to appearances, uncontrolled change is not dominant anywhere in the universe, Spenser tells us that her face becomes visible.

> Ne *Nature* to or fro spake for a space,
> But with firme eyes affixt, the ground still viewed.
> Meane while, all creatures, looking in her face,
> Expecting th'end of this so doubtfull case,
> Did hang in long suspence what would ensew,
> To whether side should fall the soveraigne place:
> At length, she looking up with chearefull view,
> The silence brake, and gave her doome in speeches few. (VI.vii.57)

Hamilton suggests that "In effect, Nature's veil is removed by the final clarity of her judgment" (734), but Hamilton's declaration that Nature makes the truth manifest seems as questionable as Nature's own assertion that change is controlled by Aristotelian teleology. An earlier passage provided a more convincing, if paradoxical, account of her visibility: "Great *Nature*, ever young yet full of eld, / Still mooving, yet unmoved from her sted; / Unseene of any, yet of all beheld" (VII.vii.13.2–4). This paradoxical formulation shows that poetry may have been, at this moment, in a better position than natural philosophy to articulate the crux that nature had become. The question of whether the face of nature is legible, however, is pervasive throughout the poem. Just as Spenser has felt compelled to repeat scenarios that rehearse the worrying question of whether the whole universe was subject to change and decline, so too does he seem compelled to repeatedly assert and question the legibility of the cosmos. Book II lets us see how issues arising from natural philosophy help to drive this compulsion and Spenser's poem.

TEMPERANCE

The proem to Book II, the book of Temperance, directly takes up the questions of whether the truth is always manifest or whether it can be hidden, whether human senses are adequate to discern the truth by reading the signs offered up by nature, and whether "new" knowledge unknown to antiquity can still be discovered.[39] The narrator worries that his poem will be judged to be "painted forgery,/ Rather then matter of iust memory" since Faery Land is unknown to his readers, a place that he can "no where show," and "which no body can know" (II.pr.1.4, 8, 9). Obviously on one level this is the case: Faery Land is a fiction, it *is* a "painted forgery" made up by Spenser. But Spenser links the plausibility of his fiction to the contemporary question of whether nature retains hidden secrets that can still be discovered by those who can correctly read their signs. He first uses the example of the discovery of the New World as proof that "of the world least part to us is red," asking "fruitfullest *Virginia* who did euer vew?" (II.pr. 2.2, 9). William Eamon has noted that the discovery of the New World created a "heightened consciousness of novelty," leading to increasing awareness that "like the New World, nature stood before investigators as uncharted territory."[40]

Spenser goes on to argue that astronomical discoveries will also reveal that the cosmos has hidden secrets that will be made manifest in the future.[41]

> Yet all these were, when no man did them know;
> Yet haue from wisest ages hidden beene:
> And later times things more vnknowne shall show.
> Why then should witlesse man so much misweene
> That nothing is, but that which he hath seene?
> What if within the Moones faire shining spheare?
> What if in euery other starre vnseene
> Of other worldes he happily should heare?
> He wonder would much more: yet such to some appeare (II.pr.3)

The question "Why then should witlesse man so much misweene/That nothing is, but that which he hath seene?" anticipates but reverses Mutabilitie's claim that "what we see not, who should us perswade." Here Spenser tries out the position, advocated by "secrets" traditions, that some truths about the world are hidden and can be found out only by special technologies or esoteric practices.

Spenser goes on to echo the language of earlier writers like Robert Recorde when he argues that Faery land can be located "By certaine signes here set in sundry place" (II.pr.4.2). These signs are necessary because readers' senses are "too blunt and bace" (II.pr.4.4) to perceive the truth directly, and they are unable to track the faint trail left by truth without "an hound" (II.pr.4.5). Eamon has shown how "the metaphor of science as a hunt occurs repeatedly in the scientific literature of the early modern period," and Spenser is clearly alluding to that metaphor here.[42] Spenser complicates the tropes familiar from natural philosophical writing by connecting the "signes" and "hound" with his allegorical method: they are what enable readers to perceive Faery land, which "no body can know" directly. His poem thus imitates nature (and Nature) by concealing its most important truths beneath a "couert vele" (II.pr.5.2), providing signs on the surface that readers need to interpret in order to gain access to the truth. However, many readers have noticed that the poem does not always provide clearly legible signs, creating a situation where the reader's interpretive struggles mirror the characters' difficulties in interpreting the characters and situations they encounter. I argue that Spenser chose to write an allegorical poem in part because it provided a way to explore the most pressing epistemological questions of his time, and the emotional energy of *The Faery Queene* is largely generated by his obsessive need to stage and restage scenes where the observed surface of the world has a complex relationship with some underlying and hidden truth.

In Book II of *The Faerie Queene*, Temperance is of course an ethical concept based on Aristotle's *Nicomachean Ethics*. But, as Michael Schoenfeldt and others have noted, it is also a medical concept based in the idea that a balance or temperament of humors grew out of and contributed to temperance in the ethical sense.[43] Schoenfeldt traces the implications of a Galenic model of temperance for self-regulation in the poem, but we have also seen that Galenic medicine was subject to the epistemological issues that beset all Aristotelian sciences in the late sixteenth century. Earlier, Thomas Elyot's *Castel of Health* (1539) offered straightforward advice about how to read signs of disease and health on the body. Elyot explains that "complexion is a combination of two diverse qualities of the iiii elementes in one body."[44] He gives the example of a "Sanguyne" complexion, in which "heat and moisture have soverayntie" and which "is perceyved and knowen by these sygnes . . . carnositie or fleshynes . . . the veynes and arteryes large" (2r). In Elyot's Galenic system, "distemperature" of the humors causes disease, "wherefore to know the distemperature these sygnes following shold be concydered" (3r).

Levinus Lemnius, a continental writer whose medical treatise was translated into English in 1576 as *The Touchstone of Complexions*, is described by his English translator as having traveled "into the verye bowels of nature" to bring back knowledge of the body.[45] Lemnius's book makes clear the extent to which humoral balance or temperament was thought to determine aspects of character, and it also shows how physical signs of humoral balance were thought to render aspects of character legible on the surface of the body. The title page promises that the book contains "most easie rules and ready tokens, whereby one may perfectly try and thoroughly know, as well the exact state, habite, disposition, and constitution of his own body outwardly," as well as "inclinations, affections, motions, and desires of his mynd inwardly."

Lemnius's account makes clear that temperament of humors is closely linked with ethical temperance or moderation when he suggests that "crasis or temperamente . . . is an agreement, and conveniencie of the first qualities and Elementes amonge themselves: Or, an equall mixture or proportion of the qualityes of the Elements, wherein no excesse blameworthy or faultye is to be founde. Contrarywyse, Intemperature is that, which is compact of the disagreeable and unlike powers of Elements and qualities, and swarveth from moderation" (32r). The "perfectly temperate" subject as described in Lemnius sounds like a catalogue of Spenserian heroes. In such men there "plentifully appeareth and is evidently descryed, humanity, gentleness, frugality, equitye, modestie, and a continent moderation of all affections" (34v). Furthermore, this temperate man "may well serve for a notable patterne of virtue" (34v). The temperate subject sounds particularly relevant for temperance as represented in Book II of *The Faerie Queene* when we learn that he is "neyther rash and heady, neyther slow and lingering" (35r).

Lemnius makes clear that a state of temperance is easily legible on the surface of the body when he identifies the "notes or markes of a body perfectly temperate": "in the outward shew, shape, and behaviour of the bodye there is evidently descryed and perceyved a comely grace and portlye dignity" (34r). He further notes that "in the countenaunce, whych is the Image of the mynde, in the eyes, which are the bewrayers and tokentellers of the inward conceiptes . . . there appeareth a kind of heroicall grace and amyablenes" (36r–v). Such a man, Redcrosse would be surprised learn, is untroubled by false dreams (37r).

On the other hand, by the end of the sixteenth century in England, a Paracelsan like Richard Bostocke had, as we have seen, denied that the body offered up easily legible signs of inner truth, arguing that diseases were caused not by imbalance of humors but by invisible "seeds," that the true causes of disease were "occultum" not "manifestum," and that diseases were cured by distilled medicines and not by tempering the balance of humors.[46] Medicines revealed themselves by "signatures," resemblances that signaled with what part of the body some natural herb or mineral shared an occult linkage. Although Spenser's concept of medical temperance remains largely within Galenic orthodoxy, he reflects considerable uncertainty about whether the "outward shew" of the body reveals its inner complexion with any accuracy. His allegory thus sometimes supplies additional signs to supplement those that occur naturally on the surface of the body. Natural signs are sometimes legible, sometimes not.

The Faerie Queene, BOOK II, AND THE SIGNS OF TEMPERANCE

Book II of *The Faerie Queene* seems at first to accord with Lemnius in suggesting that truly temperate people reveal on the surface of their bodies manifest signs of their temperate state. Guyon is initially described in terms that could come straight from Lemnius: "His carriage was full comely and upright, / His countenance demure and temperate" (II.i.6.1–2). "Comely" seems at first to function in Book II as a verbal sign of a temperate complexion, and for most of the book only temperate characters appear comely.[47] The Palmer is "A comely Palmer clad in black attire" (II.i.7.2); Medina is "a sober, sad, and comely courteous Dame" (II.ii.14.5), who "comely courted" Guyon "with meet modestie" (II.ii.15.2). Guyon addresses Medina with "comely gravitie" (II.ii.39.7), while Elissa's table manners are "unworthy of faire Ladies comely governaunce" (II.ii.35.9). In Alma's castle, Diet is "a comely personage" (II.ii.27.6) and Concoction is "full of comely guise" (II.ii.31.2). Comely appearance or behavior seems at first to be an infallible guide to temperance.

But as always in *The Faerie Queene*, this clear distinction based on appearance is questioned and eventually breaks down. Guyon's Palmer, his guide who acts with temperance more consistently than Guyon does, is initially described as "comely," "clad in blacke attire," "Of ripest yeares, and haires all hoarie gray," all signs of moderation, wisdom, prudence.[48] But the narrator comments, "if by lookes one may the mind aread, / He seemd to be a sage and sober sire" (II.i.7.2–7). Spenser seems at first to share Lemnius's assurance that the body presents "tokens" that reveal "the inclinations, affections, motions, and desires of his mynd inwardly" but is now moved to question whether "by lookes one may the mind aread." Although the Palmer never does anything that would lead us to question his wisdom and temperance, his description and the comment that "he *seemd* to be a sage and sober sire" echoes the narrator's account of the evil Archimago in Book I when he first appears to Redcrosse and Una disguised as a hermit, "in long blacke weedes yclad," "his beard all hoarie gray," "Sober he seemed, and very sagely sad" (I.i.29.2–4).[49] Although the Palmer's appearance evidently provides an accurate indication of his inner temperance, the narrator, perhaps recalling Archimago, feels compelled to question whether manifest appearance reveals hidden inner truth.

Indeed, as the book proceeds, flawed and intemperate characters are also described as "comely," and the balance shifts from mostly associating comeliness with temperate characters to mostly associating it with the intemperate: Furor's attack on Pyrochles "fowly battered his comely corse" (II.v.23.5); the bad Genius of the Bower of Blisse is "a comely personage of stature tall" (II.xii.46.4); Excesse is "a comely dame" (II.xii.55.7). By the end of Book II, comeliness is no longer a legible sign of temperance; it can also betoken its opposite, excess.

To complicate this issue of whether the body makes manifest legible signs of inner temperance, Guyon, initially described as revealing his temperate nature on his body, is for much of the book completely covered from head to toe: "all armed in harness meet,/That from his head no place appeared to his feete" (II.i.5.8–9). Knights covered by armor for battle are recognized by the emblems on their shields and not by the appearance of bodily signs. In one telling episode, Guyon, inflamed by Archimago and angrily bent on attacking Redcrosse, stops charging at the sight of Redcrosse's emblem, "the sacred badge of my Redeemers death,/Which on your shield is set for ornament. " (II.i.27.6–7).

When the Palmer (who did not hasten his temperate pace) catches up with Guyon, he immediately has "perfect cognizance" of Redcrosse's identity (II.i.31.5). Fowler glosses the word in this way: "Cognizance: recognition and understanding; in heraldry, the device by which a person is known. The Palmer instantly knows the Knight by his Red Cross and confirms Guyon's belated recognition" (176). The *OED* further tells us that a *cognizance* can figuratively mean "a badge, emblem, mark, token." Many readers will recognize this episode as a signature moment in *The Faerie Queene*: a character encounters someone or something and has trouble "reading" the truth of its identity. A companion (usually a guide figure like the Palmer or Una) is sometimes able immediately to interpret it.[50] These scenes of failed and successful interpretation have moved many critics to comment on the ways in which (in Anne Ferry's words) "*The Faerie Queene* as a whole, and especially Book I is centrally concerned with the difficulties of interpreting the appearances of things," and, as many other critics have noted, it provides contradictory and inconsistent accounts of how the world is to be rendered legible or whether it can be.[51] The narrative sometimes suggests that characters ought to be able easily to interpret what they encounter; sometimes it suggests that they can learn to do this. It sometimes suggests that only certain "adept" characters can interpret what they see, sometimes that everyone is inevitably fooled. The poem sometimes offers an allegorical surface that obviously signals what it means; sometimes it suggests that meaning is secret and hidden.

Guyon and Redcrosse are recognized by their "cognizances," by the signs on their shields rather than by their physical appearance. In a typically Spenserian redoubling, Book II seems to be trying to show that the exterior of the body provides a manifest indication of its inner state of humoral balance, but, as if unable to trust this belief, it adds on another layer of signs to the body—the religious imagery of the knights' armor that can be "read aright" by those with proper religious training. Temperance, however, plays a role in human ability to read even these artificial signs; as we saw previously, Guyon's cognizance of Redcrosse's shield was delayed by his state of wrath.

Armor is not an infallible guide to a knight's identity, however, since it can be stolen or falsified. Braggadochio, the cowardly knight who first appears in Book II and reappears periodically (along with his on-again, off-again girlfriend, the false Florimell) until his shamed exit in Book V, provides an example of Spenser's obsession with the various ways in which the body does and does not offer up legible signs of its inner truth. In Book II, Canto iii, Braggadochio enters, is clearly identified as a "losell" with "baser brest," and steals Guyon's horse and spear, providing himself with the unearned trappings of knighthood (II.iii.4). As he encounters Archimago, Belphoebe (II.iii.21–46), Sir Ferraugh (III.viii.16–19), Cambell, Blandamour, Paridell, and Satyrane (IV.iv.8–20), and Marinell (V.iii.10), his cowardly behavior repeatedly reveals his true nature. The only person he seems to fool is the false Florimell, who chooses him over other true knights (IV.iv.26).

The false Florimell also exemplifies Spenser's interest in surfaces that conceal the truth. A cyborg created by a witch out of snow, wax, golden wire, and an animating "spright," the false Florimell presents a beautiful appearance that repeatedly fools almost everyone who sees her. We are told in Book IV that "that *Florimell* her selfe in all mens vew / She seem'd to passe: so forged things do fairest shew" (IV.v.15.8–9). Although her bodily appearance can deceive everyone (except Britomart), her falsity is revealed by the girdle of the real Florimell when it (at least initially) refuses to be fastened around her waist (IV.v.16). At this point, she chooses Braggadochio over worthier knights, but despite the doubts afforded by the girdle, the others (Satyrane, Blandamour, Paridell) continue to pursue her.

Book V, Canto iii, presents several instances where "signs" or "tokens" of bodily proof are adduced in an attempt, yet again, to reveal the truth about Braggadochio and the false Florimell. Spenser seems to be trying, like Artegall and Nature, to prove without a doubt that certain knowledge can be attained from manifest signs. When Braggadochio tries to claim credit for feats of arms actually performed by Marinell wearing Braggadochio's armor, Marinell insists that the body and not its covering armor can provide the only true "signes" or "proofe" of true knightly prowess: "shew the wounds, which unto thee befell; / Or shew the sweat, with which thou diddest sway / So sharpe a battell" (V.iii.21.7–8). Marinell then reveals his own body: "and these the signes, (so shewed forth his wounds) / By which that glorie gotten doth appeare" (V.iii.22.3–4). Immediately after Braggadochio is unmasked by Marinell, the false Florimell is set alongside the true Florimell "to make paragone, / And triall, whether should the honor get" (V.iii.24.3–4). The human beings present are not allowed to attempt this distinction: as soon as the true Florimell appears, the false one "vanisht into nought: / Her snowy substance melted as with heat, / Ne of that goodly hew remayned ought, / But th'emptie girdle, which about her wast was wrought" (V.iii.24.6–9). Definitive proof of the false Florimell's false appearance is finally offered, although we learn that the previous proof (the refusal of the girdle to fasten about her waist) has been voided.

Spenser uses the word *token* as a noun in only one episode of *The Faerie Queene*, and it is not a surprise that it appears in this scene where proofs and signs are being obsessively redoubled. After the false Florimell melts away, Guyon recognizes his horse, which Braggadochio stole in Book II. Although Braggadochio's inability to "menage" the horse was noticed by the horse itself ("He had not trayned bene in chevalree. / Which well that valiant courser did discerne" [II.iii.46.5–6]) and by the narrator ("But chiefly skill to ride, semes a science / Proper to gentle bloud; some others faine / To menage steeds, as did this vaunter; but in vaine" [II.iv.1.7–9]), Guyon needs to present multiple proofs of the horse's true identity. Artegall asks Guyon if the horse has any "privie tokens" of his identity, and Guyon identifies a secret signature mark, a "blacke spot" inside the horse's mouth "shapt like a horses shoe" (V.iii.32). Like a Paracelsan signature, the sign resembles part of the thing of which it is a token: the horse's identity is revealed by a mark shaped like a horseshoe.

When two bystanders attempt to look for this secret mark, the horse kicks one in the ribs and bites the other. Guyon's true ownership is doubly revealed, then, when the horse recognizes his voice, allows him to open his mouth, "and suffred all his secret marke to see" (V.iii.34.6). Providing further signs that he recognizes Guyon, the horse "follow'd him with gladfull glee, / And friskt, and flong aloft, and louted low on knee" (V.iii.34.8–9). After the multiple signs and tokens, Artegall "did plaine arreed, / That unto him the horse belong'd," and he orders Guyon to ride off on the horse while "that losell, plainely now displayd, Hence fare on foot" (V.iii.35.1–2, 5–6). When Braggadochio protests, Artegall has his false signs of knighthood destroyed; "then from him reft his shield, and it renverst, / And blotted out his armes with falsehood blent, / And himself baffuld, and his armes unherst, / And broke his sword in twaine, and all his armour sperst" (V.iii.37.6–9). Braggadochio has been "displayd" as a "losell" from his first appearance in the poem. Guyon's true ownership of the horse has never really been in question. At this point in the poem, though, the question of whether any sign, whether bodily or manmade, whether manifest or occult, can reveal the truth is so fraught that the signs must be doubled and redoubled, and then destroyed with equal fervor.

Although the *Mutabilitie Cantos* seem to suggest that the surface of nature is unreadable because nature chooses to hide itself from human sight, the proem to Book II suggests that the fault may also lie in the failure of human senses, which are "too blunt and bace" to perceive the truth. Book II further explores this possibility in its allegorical centerpiece, the Castle of Alma, where Arthur and Guyon are allowed to journey into the body to observe its hidden inner processes.

INTO THE BOWELS OF NATURE

The Castle of Alma has often been criticized as a particularly clumsy bit of allegory. Gordon Teskey calls it one of "the most simplistic allegorical episodes, wherein the teeth bow to the soul," and Lauren Silberman has described it as "awkward and problematic in execution."[52] Yet it provides insight into how Spenser's view of the body and the adequacy of the senses was shaped by the naturalistic and secrets traditions. The whole episode picks up on the common trope of a journey to a realm that is otherwise hidden and inaccessible—in this case the interior of a living body. Michael Schoenfeldt links Spenser's castle of the body with Elyot's *Castel of Health*, noting that, in the Galenic system, "psychological, ethical, and physiological health is an edifice perpetually being constructed and in need of continual maintenance."[53]

Although Guyon and Arthur's visit to the castle emphasizes, as Schoenfeldt has shown, the bodily operations of digestion and concoction central to Galenic regimens of health, their entrance is prefaced by a stanza that focuses on the body's mathematical and cosmological "frame" in terms more suited to the tradition of Neoplatonic and Hermetic secrets. Here Spenser calls on what Dee, in the preface to Euclid, describes as "Anthropographie," or "the description of the Number, Measure, Waight, figure, Situation, and colour of every diverse thing, conteyned in the perfect body of MAN: with certain knowledge of the Symmetrie, figure, waight, Characterization, and due local motion, of any parcel of the sayd body, assigned: and of Numbers, to the sayd parcel appertaining" (C4r). Spenser first describes the body in terms of symbolic geometry: "the frame thereof seemd partly circulare, / And part triangulare, O worke divine . . . And twixt them both a quadrate was the base" (II.ix.22.1–2, 6).[54] He also identifies the body with the numbers seven and nine, relating the microcosm to the number of planets and Ptolemaic spheres. The surface of this body does not reveal itself by humoral symptoms but through a complex set of esoteric mathematical symbols.

Once they enter the castle, however, the allegory becomes more straightforward and, in the opinion of many critics, simple-minded: the interior of the Galenic body is easily legible. They enter the mouth, go down the throat into the stomach, watch digestion in action, then ascend to the heart and on to the brain. Unlike Mutabilitie's journey into the heavens, Guyon and Arthur's fantastic voyage provides them with factual information about the inner workings of the body, albeit information that was readily available in many Galenic texts without having to imagine such a voyage.

After pausing in the heart to meet the affections, the two heroes climb up the "ten steps of Alabaster wrought" that represent the spinal column to visit the brain/turret, seat of cognition (II.ix.44.9). We learn that the brain has "diverse rooms, and diverse stages," but there are three "chiefest" rooms in which dwell "three honorable sages" (II.ix.47). Spenser's allegory breaks down here, and critics have disagreed about what the sages represent. One school of thought has argued that they represent "the three interior senses of the mind or higher faculties of the sensitive soul."[55] Harry F. Berger and Robert L. Reid accord with this idea that the three sages do not "represent the highest power of intellect," and Berger insists that they are confined to "sensitive" rather than "intellectual" functions.[56] Lynette Black, on the other hand, argues that they do represent the intellectual faculty of prudence and function as "the mind of the rational soul."[57]

Spenser's representation of the three sages seems to conflate two slightly different traditions: an iconographic depiction of a tripartite prudence, which is based on knowledge of future, present, and past, and a Galenic account of faculty psychology which locates human cognitive function (including reason) in the brain.[58] Elyot's *Castel of Health* identified these three parts in the Galenic tradition as "Imagination in the forhed," "Reason in the brayne," and "Remembrance in the nodel [back of the head]" (11r), and Spenser's account reflects this spatial configuration. The sages of the first and third chambers are clearly part of this Galenic tradition and are explicitly identified as imagination (*Phantastes*) and memory (*Eumnestes*).[59]

The middle sage, who should represent reason, remains unidentified. Berger notes that the "middle chamber has caused some bewilderment" and that the middle sage's "precise function is not clear."[60] As critics have noted, reason is a crucial factor in temperate behavior. Helen Cooney identified the Palmer with "right reason" or "prudence," and Lynette Black has argued that prudence "permeates the structure and meaning of the quest of Sir Guyon."[61] So why would Spenser refrain from assigning this sage a name? Why does Alma hustle Guyon and Arthur out of his chamber despite their desire to linger there?[62] Why is there uncertainty (as evidenced by critical disagreement) about the extent to which a truly "intellectual" or "rational" power is even present in the turret and therefore in the castle as a whole?

A closer look at the middle sage reveals that he shares qualities with experience in the Aristotelian sense. He judges the evidence of the senses in accordance with established bodies of knowledge, and through "continuall practice and usage" has "growne right wise, and wondrous sage" (II.ix.54.4, 5). His pivotal place in the brain and Alma's reluctance to linger or to explain who he is may reflect his conflicted situation in late-sixteenth-century England. For Spenser, the faculty of Reason responsible for judging and making sense of the data provided by the senses had been called into question by the contemporary loss of confidence that the manifest appearance of the world as perceived by the senses provided access to the truth.

The middle sage sits in a chamber that contains images that represent a wide variety of human bodies of knowledge about the world. This includes knowledge of legal practice and theory ("Of Magistrates, of courts, of tribunals," "Of laws, of iudgements, and of decretals"), knowledge of politics ("of commen wealthes, of states, of policy"), and, finally, the rest of human knowledge: "all artes, all science, all Philosophy,/And all that in the world was aye thought wittily" (II.ix.53). As Berger has noted, the middle sage deals in images and not words because sense impressions take that form.[63] The sage meditates on these images and has attained wisdome "through continuall practice and usage" (II.ix.54.4). When we recall that Aristotelian naturalism was based on the assumption that ordinary repeated experience of the world provided by the senses gave access to truth about its nature, we can see the resemblance between this sage and Aristotelian experience. Gabriel Harvey's marginalia may provide a useful gloss on the middle sage: "All sciences are based on the senses and reason, clearly they accord with reason, and with sense. Reason is the soul of each of the principles. Experience is the soul of souls, the firmest demonstration and irrefutable criterion. Give me visible and rooted demonstration of each principle, experiment, geometric instrument, astronomical, cosmographic, horologiographic, geographic, hydrographic, or mathematical in any way."[64]

Before we hear about the sage's immersion in orthodox bodies of knowledge and the manifest evidence of the senses, we learn that his chamber contains traces of "secret" traditions in its accounts of "the memorable gestes,/Of famous Wisards" (II.ix.53.3–4). It's possible that this alludes to something like the "The Famous Historie of Fryer Bacon," or the *English Faust Book,* which recount the fanciful acts of magic and conjuring performed by scholars believed to have rejected the orthodox disciplines that the sage will meditate on in favor of esoteric "occult" (in the spooky sense) practices. However, this kind of magic is denigrated throughout the poem (in figures like Archimago and the witch who creates false Florimell), so it seems more likely that these "wisards" are learned men (like the ones praised by Gabriel Harvey) who gain knowledge not from continuous ordinary experience but instead from "experimenta," specific acts (or "gestes" in Spenser's poetic vocabulary) carried out with scientific instruments and designed to provide access to discrete, hidden facts. If so, Spenser's middle sage is straddling an epistemological divide. Like Spenser's heroes throughout the poem, the middle sage is torn between a belief that knowledge about the world is manifest and a belief that it is occult; that the senses are trustworthy and that they are not. That wisards and their gestes precede orthodox knowledge may mirror the order of the sages, which proceed from foresight, to the present, then to memory.

This is why, despite Guyon and Arthur's desire to become "disciples" of the middle sage and to continue their "great pleasure" in "his goodly reason, and grave personage" (II.ix.54.6–7), Alma leads them on to the next chamber. The middle sage is glossed over so that the knights can move on to Memory, so that they can learn from the written documents of history rather than from observation of a natural world that has become too problematic to yield sure knowledge. Memory, seemingly unlike the rest of the universe, is impervious to change. His records are contained in an "immortal scrine,/Where they for ever incorrupted dweld" (II.ix.56.6–7). Memory's scrine may be the one source that Mutabilitie cannot control, since the narrator tells us that he is able to recount Mutabilitie's lineage because he "found it registered of old,/In *Faerie Land* mongst records permanent" (VII.vi.2.3–4). In this light, we can now notice that the middle sage himself does not directly judge his experience of nature but rather ponders the authoritative bodies of knowledge that describe it for him. Spenser's middle sage thus seems to embody the commonplace seventeenth-century criticism that sixteenth-century science was based on authoritative texts rather than observation of nature.

In Alma's turret, Spenser, as he does throughout the poem, values historical narrative over scientific description. If the middle sage marks the place where human cognition ought to be able to accurately interpret the data provided by the senses, then the whole poem stands as a monument to his failure. But the sage's reliance on authority rather than experience appears in the context of a poem that repeatedly stages interpretive encounters with experiential knowledge and that seems urgently concerned with whether, and under what conditions, the world reveals the truth about itself. Gordon Teskey has argued that Spenser is "the greatest poet of appearances, of manifestations," even as he "harbors a distrust of his own as well as of others' power to make manifest."[65] This contradiction emerges at least in part from late-sixteenth century uncertainty about whether, and how, the world of nature manifested itself to human subjects, and this uncertainty is one of the driving engines of Spenser's poem.

CHAPTER 5

Shakespeare and New Forms of Nothing

Spenser expressed his engagement with disturbing new ideas only indirectly in *The Faerie Queene*. Still, it seems that late sixteenth-century writers considered epic poetry a more suitable genre for the exploration of cosmology than lyric or dramatic works. Dramatic writers seem to have been particularly unwilling directly to represent new ideas and technologies onstage. Take, for example, the experiments in optics leading up to the invention of the telescope. Several plays do incorporate optical devices, but they show them to be a product of demonic magic (*Friar Bacon and Friar Bungay*), an analogue of the illusionistic power of the stage (*Travels of the Three English Brothers*), or a fraudulent trick (*Albumazar*).[1] If we're going to look for signs of the new science in Shakespeare's plays, we should not expect to see them dramatized explicitly.

Like all poets, Shakespeare has a basic model of the world and how it works, based in intuitive experience of nature and contemporary natural philosophy. His imagery is shaped by and reflects this mental model. Gail Kern Paster, for instance, has shown how humoral physiology was fundamental to representations of the body in early modern drama. The language and imagery of these plays is imbued with humoralism, and their representations of human behavior are inflected by this set of assumptions about the nature of embodiment.[2] Shakespeare's understanding of the natural world is just as clearly based on a version of Aristotelian naturalism; the four elements and the tenets of meteorology subtend his descriptions of the world and its workings.[3] New ideas that challenge these basic Aristotelian assumptions, especially those related to the abstraction of Arabic numerals and corpuscular theories of matter, surface in his sonnets and in *King Lear*. The sonnets reflect the promise and danger of a newly abstract mathematics, and the bleak vision of *Lear* is at least partly a response to a frightening new world in which the fabric of the universe is riddled with nothingness.

Shakespeare does not seem to have been particularly interested in astronomy. There are no references to Copernicus, to the new star, to precession, or to the issue of supralunar change in his works.[4] A number of scholars have noted Shakespeare's interest in changing ideas about mathematics, especially the introduction of Arabic numerals and the "cipher" or figure for zero, and have persuasively shown how these concepts were bound up with new technologies of commerce, such as double-entry bookkeeping, that were important in the development of a commercial theater.[5] Shankar Raman has argued that advances in algebra during the period may be related to Shakespeare's fascination with numbers in several plays.[6] There is also evidence that Shakespeare was interested in changing ideas about the nature of matter, especially concepts such as divisibility and void space that arose as particulate theories challenged the Aristotelian elements.[7] While Spenser's *The Faerie Queene* seems concerned with exploring whether new ideas threatening the stability and intelligibility of the universe were actually true, Shakespeare explores the existential implications of some of those ideas.

THE SONNETS AND ABSTRACT NUMBERS

When Shakespeare sat down to write his sonnets, he seems not to have recalled (or not to have cared about) Sir Philip Sidney's criticism of earlier sonnet writing in his *Defence of Poesy*. Sidney remarked that "many such writings as come under the banner of unresistable love" would never persuade their addressee that "they were in love: so coldly they apply fiery speeches."[8] Rather than love, Shakespeare's sonnets are preoccupied with the passage of time and the economics of value, and they are, for lyric poems, surprisingly abstract. The sonnets' economic language and preoccupations have been discussed by many critics, including Thomas Greene, Natasha Korda, Lars Engle, and Peter Herman. Engle and Herman, as well as Heather Dubrow and others, have noticed that the sonnets tend to dwell on the general rather than the specific or particular.[9] I argue that these aspects of the sonnets are related: their preoccupation with economics is grounded in an interest in mathematics, and Shakespeare's interest in the mathematical is prompted in part by a pervasive process of abstraction that altered mathematical knowledge in western Europe in the last half of the sixteenth century after Arabic numerals, place value, and zero were introduced. Mathematics began to offer powerful new ways to think abstractly, and Shakespeare's sonnets register the appeal of this new way of thinking, as well as the anxieties that it caused, while *King Lear* pursues only the anxieties.

Christopher Martin's sense that the procreation sonnets display a "shallowness of conception that dooms them to . . . facile posturing" or his belief that the speaker "would just as soon stick to the homey blur of abstracted tradition"[10] is not likely to be mitigated by the argument that they're really all about arithmetic. But the changes in mathematical theory in this period were related to profound changes in how people conceived of the universe and their place in it. I believe that Shakespeare was aware of these changes and was both excited and disturbed by them. If we know more about what was happening to numbers, we can better understand why they have such a hold on Shakespeare's imaginative vision in these poems.

Three different sonnets allude to one of the foundational concepts of ancient mathematics, the idea that one is not a number. In the concluding line of Sonnet 8, Shakespeare uses this idea in its proverbial sense to suggest that if the friend remains unmarried, he will leave nothing of himself behind when he dies: "thou single wilt prove none." Sonnets 135 and 136 also allude to this idea that "one is reckoned none" (136) as the speaker attempts to calculate the multiplicity of the lady's lovers.[11] Understanding why ancient mathematicians did not consider one to be a number and tracing the struggles of Robert Recorde to acknowledge and go beyond this concept can help us see why Shakespeare was so fascinated with this and other changing mathematical theories.

As noted in chapters 2 and 3, ancient Greek (and Roman) mathematics was, in some respects, like much ancient science, deeply essentialist: that is, all valid knowledge about things in the world was thought to be knowledge about their essential nature. Focusing on Galileo's theories of motion, Margaret Osler has argued that the scientific revolution involved "a shift from an epistemological position which may be called essentialism to a position of nonessentialism," and she describes ancient essentialism as "the assumption that either the world in some way corresponds to our conceptions or that our observations in some manner reveal the inner nature of objects to us." Essentialism is thus connected with intuitive knowledge of the world and entails the belief that ordinary observations reveal not just superficial knowledge but access to the essence of things. The new science, however, "came to deny the possibility of any knowledge of essences."[12] The basic arithmetic and geometry inherited from antiquity were both bound by rules designed to maintain a connection between the nature of objects in the world and the ways in which they could be counted or measured. Ancient arithmetic was by nature tied to its role in enumerating objects. We can recall Reviel Netz's argument that the pervasive use of counters for practical calculation in antiquity had a profound effect on mathematics and ancient culture in general.[13]

As we saw in chapter 2, one was not considered to be a number because it defines the kind of thing that is being counted.[14] Only more than one constitutes a multiplicity of things and, therefore, a number. One was also thought to be indivisible, since in counting you enumerate discrete, whole objects, and the version of arithmetic theory inherited through Boethius from antiquity dealt only with whole numbers and did not believe that fractions could exist.[15] Emphasis on the nature of things being counted meant that different kinds of things were thought to be incommensurable and could not be included in the same equation. This incommensurability rule proved to be a barrier to progress in the physics of motion until, in the fifteenth and sixteenth centuries, people like Galileo began to include distance and time in the same equation.[16] Algebra existed only in a rudimentary form because it was difficult for ancient mathematicians to let variables stand in for numbers, since a variable was completely abstracted from any specific kind of thing.[17] All of these rules meant that the mathematics early modern Europe inherited from classical antiquity was tied to the identity of objects and to the concrete world.

Various ancient mathematicians eroded, evaded, or broke these rules, but they persisted, in rudimentary form, into the Middle Ages, while more sophisticated mathematical ideas were lost.[18] Calculation during this period was primarily carried out with counters, and numbers were represented by Roman numerals. It took the separate mathematical traditions that arose in India and the Islamic world during the Middle Ages to break some of the inherent limitations of ancient mathematics. These alternate mathematical systems, which included the use of Arabic numerals and zero, took hundreds of years to achieve general adoption in Western Europe. The fifteenth and sixteenth centuries saw European advances in mathematical knowledge and further dissemination of this new knowledge, in part because increased trade required more complex accounting practices than could be carried out using Roman numerals and counters.[19] Merchants began to use a kind of basic algebra called the "Rule of three," or "golden rule," to figure out unknowns involving more than one kind of thing in the same equation.[20]

In England, Robert Recorde's extremely popular book on basic arithmetic called *The Grounde of Artes* (1544) introduced operations with Arabic numerals and went through twenty-nine editions. He also published a sequel, *The Whetstone of Witte* (1557), which offers an introduction to the basic cossic algebra developed by merchants. As we saw in chapter 3, when Recorde introduces the fundamentals of addition using Arabic numerals in *The Grounde of Artes,* he must explain three new concepts: "the Value, the Figure, and the Place." He introduces the figure that "doth signifie nothing, which is made like an o," and also the nine digits, which are "signifying figures," each having "two values: one alwayes certaine that it signifieth properly, which it hath of his forme: and the other uncertain, which he taketh of his place."[21] Already numbers, including the number one, are seen as signifiers with multiple meanings rather than being linked to the essential identity of a thing. Still, Recorde here maintains the rule that you should not "adde together two summes of sundrie thinges, which you ought not to do, except you seeke only the number of them, and care not for the things" (50). Later in the book, however, he introduces the merchant's "golden rule" to determine an unknown term, and here he does include "sundrie things" in the same equation, for example "if the cariage of 1 C weight (that is to wit 112 pound) 30 miles doe cost 12 pence, what will the cariage of 5 C weight cost being caried 100 miles" (199). Here weight and distance are considered commensurable and are included in the same equation.

The Whetstone of Witte moves farther into the realm of the abstract while still grappling with the specter of ancient rules. It begins with doggerel that takes up and fudges the issue of whether one is a number:

> One thyng is nothyng, the proverbe is
> Which in some cases doth not misse.
> Yet here by woorkyng with one thyng,
> Soche knowledge doeth from one roote spryng,
> That one thing maie with right good skylle,
> Compare with alle thing.[22]

I doubt that anyone has ever put this poem in the same equation with Shakespeare's sonnets, and they are in many ways incommensurable. And yet, Recorde plays on the number one, identity, multiplicity, and abstraction, as Shakespeare does in Sonnets 36 and 136, which also turn on these concepts.

At the beginning of *The Whetstone*, Recorde struggles with the ancient dictum that one could not be divided. He notes that neither Euclid nor Boethius allow fractions, although Recorde had taught how to perform basic operations on them in *The Ground of Artes*. He explains that fractions were disallowed "bicause all nombers doe consiste of a multitude of unities: and every proper fraction is lesse than an unitie, and therefore can not fractions exactly be called nombers" (A2v). He concludes that "hereby I se that 1 is not to be called a nomber for then all nombers above it, must needes be *compounde*, because thei consist all of unities" (A3r–v).

In *The Whetstone*, Recorde also struggles with the relationship between number and thing, identifying two different kinds of number: abstract and contract. A contract number departs "from his generall libertie of signification" and is "bonde to one denomination, as in saiying 10 grotes (where 10 is restrained from the liberty of valowyng any other thing but grotes)." Abstract numbers, on the other hand, are "when I saie 10 and doe not limite any denomination, then is that 10 abstracte and severed from all specialities, and standeth free to any name of thing" (A2r). Abstraction offers freedom but perhaps at the cost of a direct connection between number and thing.

This movement toward abstraction in mathematics is, of course, closely connected with the abstraction that has been linked to the beginnings of capitalism. Merchants needed more abstract mathematics to deal with an expanding global system of trade. Marxist analyses have described the "increasingly impersonal, disembodied, and abstracted character of early modern exchange."[23] Mark Netzloff has emphasized the increasing reliance on "forms of representative currency, such as credit or bills of exchange," which rendered money itself more abstract.[24] There were a number of reasons to condemn usury in this period, but among them was a concern that if money could reproduce itself, then it became abstracted from labor and the exchange of concrete goods. Knowledge of newly abstract mathematics was increasingly important to merchants and tradesmen as it became a more flexible tool for calculation. Linda Woodbridge has established that there was a proliferation of bookkeeping schools and of published math books in this period; in her words, "mathematical publishing intensified after the Armada attack."[25] This new mathematical knowledge was thus widespread, and we cannot separate the spheres of economics and mathematics in this period.

Thomas Greene compellingly traced the "cosmic and existential economics that haunts the sonnets," arguing that the sonnets express "a bourgeois desire to balance cosmic and human budgets" which is "thwarted by a radical flaw in the universe, in emotion, in value and in language."[26] Mathematical abstraction is deeply implicated in this project, since it enables unfettered accumulation, but it disconnects that accumulation from material reality. The procreation sonnets provide a striking example of the way in which abstraction can both liberate and undermine.

Critics have long puzzled over (and variously attempted to explain) why Shakespeare was the first poet to dwell on immortality through procreation and why he chooses to dwell on it at such length.[27] Biology doesn't explain this. In fact, it makes the idea less plausible, since early modern theories of procreation are conflicted about the role of each parent in conception and about what is inherited from each. Although most of these theories give a larger role in procreation to the father, and Shakespeare seems to have shared this view, the womb was thought to have a shaping influence on the nature of the child, and the maternal imagination could also intervene, bringing about an offspring with no resemblance to either parent.[28] The sonnets themselves are fairly vague about what the son actually inherits from the father, variously mentioning "memory," "blood," and "shape." In the absence of DNA evidence, no father could be sure a child was his, an idea that causes great anxiety in Shakespeare's plays. The idea that fathering a child could provide a reassuring bulwark against mortality seems, in fact, quite odd if the father can never know for sure whether or not the child is his. So why does Shakespeare so obsessively explore this idea in the first seventeen sonnets?

The ongoing abstraction of mathematics, including the recognition that one is a number and that "abstract" numbers can be separated from the identity of the things being counted, contributes to the speaker's ability to generalize from father to son in these sonnets. Several critics have noted the slippage from general to specific in the opening lines of Sonnet 1: "From fairest creatures we desire increase, / That thereby beauty's rose might never die," addresses the kind of abstract immortality available to a species, not an individual (ll. 1–2). Lars Engle aptly describes these lines as more appropriate, in their abstraction, for a voice-over in a Sierra Club commercial.[29] The next two lines attempt to apply this concept of immortality through reproduction to a specific individual: "But as the riper should by time decease / His tender heir might bear his memory." Critics have questioned whether this slippage from abstract to particular is truly convincing, but it is thinkable only if "increase" is conceived of as "abstract"—"free to any name of thing"—rather than "contract"—tied to a specific individual.[30] This is why the friend who is "contracted to thine own bright eyes" (l. 5) remains a one that is not a number and in its unity cannot be divided. Only if the "one" attached to the father is conceived of as abstract can his "increase" in the form of a son be commensurable with himself. Otherwise, father and son are two different individuals and cannot be added together. Therefore, according to traditional mathematics, if the father were considered a unique individual, then his son could not "sum" the account of the father as in Sonnet 2, where "this fair child of mine / Shall sum my count, and make my old excuse'— / Proving his beauty by succession thine" (ll. 10–12). If one is in fact a number, it can increase itself without losing its identity. If "sundry things" can be contained in the same equation, then the son is able to "sum" the father's count.

Peter Herman has pointed to the problematic effect of seemingly positive references to usury in the procreation sonnets, since references to usury in Shakespeare's plays are always negative, and usury was most often condemned in the period as connoting "all that was destabilizing and threatening in the socioeconomic developments affecting early modern England."[31] Herman argues that "the poems' homology between usury and procreation creates shifting and often incoherent meanings."[32] I think that the contradictory valences of usury in the sonnets make more sense in the context of a new mathematics of abstraction. In Sonnet 6, for example, Shakespeare entertains the idea that abstract and exponential increase could be good: if the friend were to produce a son with someone who would "pay the willing loan," he would be "ten times happier be it ten for one. / Ten times thyself were happier than thou art, / If ten of thine ten times refigured be" (ll. 6–9). Stephen Booth notes that "refigured" means "duplicated (with a play on 'figure' meaning a numerical symbol, like 10 or 1)."[33] Although Shakespeare here imagines an interest rate ten times the 10 percent that was allowed, this unimaginable projection is entertained as the logical consequence of allowing one to be "refigured," or duplicated, but also—as Recorde teaches must be the case with Arabic numerals and place value—of assuming different meanings in different positions. Raman has noted Leontes' similar reference to the multiplicative power of zero.[34] Usury is not acceptable in the plays, where the social implications of its abstraction of value are necessarily involved. The sonnets, on the other hand, employ usury as an image that is already abstracted from social context and, in the case of the young man, in a situation where numerical abstraction is generally imagined as enabling. The procreation sonnets explore the possibilities and implications of imagining numbers in this newly abstract sense. Usury becomes thinkable in this context, although it also, as Herman argues, makes manifest the drawbacks of unlinking numbers from things.

The sonnets to the lady manifest and expand upon the problems that can result if one has the same status as two. Sonnets like 135 and 136—the famous "Will" sonnets that play on his mistress's promiscuity—take abstraction into more explicitly dangerous territory. Natasha Korda has argued that some of the anxiety in these sonnets can be linked to the prevalence of female usurers in England in the period, and Patricia Parker has noted that arithmetic was often associated, in negative terms, with early modern female sexuality.[35] These sonnets suggest that once "one" is decoupled from the identity of a specific individual, it doesn't matter who "one" is: "Among a number one is reckoned none. / Then in that number let me pass untold, / Though in they store's account I one must be, / For nothing hold me."[36] Thus, she can "Think all but one, and me in that one will."[37] The exponential multiplication that abstraction allows, the interchangeability of one, two, and zero, and the commensurability of unlike things all underlie the speaker's angry attempts to determine his place in relation to the lady and her "abundance" of lovers. That almost all of these words have doubled meanings further underscores the idea that a single unit does not have an inherent, indivisible unity, nor does it have a fixed identity.[38]

Shakespeare's sonnets are certainly "about" much more than changing mathematical theories. These ideas about number and abstraction by no means "sum their count." But I do believe that the play on mathematical concepts in the sonnets is not just superficial wordplay. Helen Vendler, who is perhaps one of the staunchest defenders of the sonnets as successful lyrics, tends to dismiss the passages that are particularly concerned with number. About Sonnet 6 she says, "In this rather labored conceit of interest-bearing funds, deliberately situated in the tenth line—on a posterity of ten producing a posterity of ten times that number reveals the degree to which Shakespeare could be entranced by fancifulness."[39] I would argue that Shakespeare's play with number goes deeper than fancy and that the new possibilities for combination and recombination and for the multiplication of difference that mathematical theory was introducing in late sixteenth century England had a powerful effect on Shakespeare's imagination. Ideas about numbers lie close to the heart of these strange and wonderful poems.

THE PHYSICS OF KING LEAR

By the time that Shakespeare came to write *King Lear*, the new math, especially as it related to new ideas about the structure of matter, seems to have become almost entirely threatening to him.[40] The possibilities of the cipher were perhaps more disturbing when nothingness was part of the fabric of the world. That *King Lear* engages with "materialism," containing, in Richard Halpern's words, "an embedded set of more-or-less 'materialist' discourses," has been noted by many critics, who have tended to read this materialism either in Marxist terms, as the replacement of a spiritually informed feudalism by capitalist materialism, or in religious terms, as hinting at the replacement of orthodox religious belief with an atheist materialism.[41] Fewer critics have taken "materialism" literally, in order to chart the implications for the play of changing theories of the nature of matter and the structure of the material universe in the period.[42] A notable exception is Henry Turner, who, in a brilliant essay on the spatialities of the play in performance and on the printed page, suggests that at some points the play "gestures toward the mathematical and spatial concept of infinity" so that "the 'nothing' of the storm could be said to perform the quasi-scientific space of the 'vacuum' or 'void,' concepts debated in both natural philosophy and in the newer Stoicism and neo-Platonism of Campanella, Bruno, Francesco Patrizi, and many others."[43]

I think it is possible to provide a more precise context for the play's pervasive sense that there is nothing beyond matter, nothing that orders or shapes it, and that its nature is uncertain and unknowable. Early modern debates about the nature of matter raised questions about whether or not it is infinitely divisible, whether substance itself depends on divisibility, whether a void or vacuum exists, whether nature was created *ex nihilo,* and whether or not a materialist atomism is necessary to explain phenomena like evaporation, condensation, and weight. All of these, I argue, are echoed in the language and imagery of the play and significantly structure its treatment of epistemological and ethical questions. I even suggest a contemporary English *locus* for such debates, in the circle of Henry Percy, Earl of Northumberland, and followers like Thomas Harriot and William Warner, whose speculations about atomism were widely known and who were, in 1606, carrying on experiments in the Tower of London, where Percy had been imprisoned on suspicion of involvement in the gunpowder plot.[44] Atomism offered a dramatic departure from previous views of matter in two ways, threatening its stability and intelligibility. First, it potentially yielded a picture of even the densest substance as discontinuous, riddled with invisible holes, full of tiny voids that threatened its coherence with the possibility of infinite divisibility. Second, it also suggested that the nature of the material world might be radically different from human embodied—and intuitive—experience of its solidity and weight.

In *King Lear,* Shakespeare responds to a growing realization that the structures developed by existing systems of thought to explain the invisible and counterintuitive workings of matter no longer work. The play marks a significant rupture between the commonsense bedrock of observed experience and existing explanatory systems. And because Shakespeare was a poet, his complex engagement with images of the physical world also structures his understanding of social and ethical problems. In *King Lear,* images of divisibility, weight, and smell can help us trace out Shakespeare's problematic physics of existence.

In the late sixteenth and early seventeenth centuries, natural philosophers began to seek a material explanation for persistent problems in physics. As John Henry has argued, "when philosophers began to recognize the need to eschew 'occult' qualities and powers from their explanations of natural systems, they turned to the mechanical philosophy which relied on the notion of contact action between bodies in motion for its explanatory force."[45] Such an explanatory system required a theory that matter was composed of tiny particles ("atoms," corpuscles," "minima," or "seeds"), since "chemical, physiological, and optical phenomena . . . can only be explained in terms of the motions and impacts of invisibly small particles." Different thinkers—Galileo, Kepler, Harriot, and later Kenelm Digby and Robert Boyle—constructed different versions of a particle theory of matter. Such theories were considered dangerous, as Henry has argued, because they verged on a potentially atheistic materialism. Many such theories asserted that the small particles of matter were indivisible and indestructible, that they had always existed, and that God could therefore not have created the universe ex nihilo. Such theories also often asserted the existence of void space between the particles, in defiance of Aristotle's rule that a vacuum could not exist in nature. Digby, a mid-seventeenth-century atomist, clearly presents the challenge to human experience of the material world entailed by the assumption that it is composed of tiny particles: "remembering how the nature of Quantity is Divisibility; and considering that rare thinges are more divisible than dense ones; we must needes acknowledge that the nature of Quantity is some way more perfectly in thinges that are rare, then in those that are dense. On the other side, more compacted and dense thinges, may happily seeme to some to have more Quantity then those that are rare; and that it is but shruncke together."[46]

Cognitive science, however, lets us see further disruptive implications of atomism in the early seventeenth century, such as the implication that it provided explanations of material phenomena based on structural analogy instead of identity. Dedre Gentner and Michael Jeziorski have demonstrated the important difference between the analogies used in modern science (for example, that electricity behaves in some ways like flowing water or that an atom resembles the solar system), which convey structural relations, and earlier use of metaphor by alchemists working with Aristotelian and Neoplatonic paradigms.[47] A structural analogy represents similar structural relations between two systems, but it does not imply similarity of attributes (Aristotelian qualities) or causal relationships (such as the "occult" causes posited through astrology or alchemy between microcosm and macrocosm).[48] Unlike the assumption behind the secret traditions like Paracelsanism that connections between what could be seen and what couldn't be seen were real and had causal force, proponents of atomism were implicitly suggesting that what you can see only corresponds structurally—perhaps even obliquely and indirectly—with what can't be seen. Atomism offered a theory of matter based on analogy with the movements of visible bodies, but it did not necessarily suggest that the atoms were actually *like* visible matter. This theory is thus disruptive not just because belief in a transcendental ordering system of spirit is gone but because the relationship between what you can see and what you can't has been questioned.

A poet or playwright, of course, isn't seeking scientific explanations for the behavior of matter in the world. Rather, he or she uses language to create a representation of the world, or part of it. To the extent that a material world is represented in a play such as *King Lear,* its representation is necessarily subtended by the playwright's intuitive model of the real world. The imagery of a play or poem will reflect its author's mental models—for example, the assumption in *King Lear* that Gloucester's "heavy substance" would have fallen from the cliff with greater velocity and more destructive force than "goss'mer, feathers, air."[49] But a poet is also in the business of explaining the invisible through analogy with the concrete and visible—so, for instance, Shakespeare in this play persistently describes love as being "heavy." *King Lear* represents a world where the nature of matter and the relationship between visible and tangible matter and qualities that can't be seen have been called into question.

The unpublished writings of Thomas Harriot provide an example of the kind of atomistic speculations that were disrupting the human relationship with the material world around 1606. Shakespeare would almost certainly not have read Harriot's writings or have known his theories of atomism in any detail. But the general outlines of Harriot's subversive thought, his reputation for atheism, atomism, and belief in creation ex nihilo, were well known at the time, through the testimony made public during the trials of his patrons Sir Walter Raleigh and Henry Percy, Earl of Northumberland, and as general gossip, reflected in John Aubrey's posthumous biography.[50] Harriot was perhaps most widely associated with the shocking phrase "nihil ex nihilo fit," as John Aubrey records: "Mr. Thomas Hariot . . . did not like (or valued not) the old story of the Creation of the World. He could not believe the old position he would say 'Nothing is made out of nothing.' But, said Mr. Hagar, a nothing killed him at last; for in the top of his nose came a little red speck (exceeding small) which grew bigger and bigger, and at last killed him."[51] Editors usually gloss Lear's "Nothing will come of nothing" as "an Aristotelian maxim," "echoing the famous Aristotelian doctrine *Ex nihilo nihil fit* (denied by Christian philosophers in respect of the Creation)."[52] However, in 1606 the phrase would have been more currently associated with Harriot's atomism and the idea that the universe was created out of preexisting atoms. Shakespeare's references to "germains" or seeds of matter (in *Macbeth*, 4.1.59; in *King Lear*, 3.2.8) and to the Aristotelian doctrine of the impossibility of a vacuum in *Antony and Cleopatra* (2.2.216), as well as the "atomies" mentioned in *As You Like It* (3.2.230–31; 3.5.12–13) and *Romeo and Juliet* (1.4.57), demonstrate at least some awareness of corpuscular theory and the controversies surrounding it in the early seventeenth century.[53]

Atomism solved some problems in understanding the behavior of matter but raised others: the existence and nature of infinity, the existence of a void or vacuum, the gap between mathematical theory and material phenomena, the gap between what could be seen and what could not. As we saw in his letters to Kepler, Harriot was drawn to the same problematic phenomena that attracted other contemporary scientists to those aspects of the physical universe which could not be adequately explained by the Aristotelian elements, including the behavior of light, the divisibility of matter, and matter's ability to change its density and weight. Divisibility was an issue because Aristotelian theory did not offer an adequate answer to the question of whether a substance could be infinitely divided into ever smaller parts that still retained the qualities of the substance or whether there was a limit of divisibility, beyond which the substance was no longer itself.[54] The contradictions resulting from the assumption that continuous bodies can be infinitely divided until the parts are invisible, or that separate geometric points somehow form a continuous line, had puzzled mathematicians since Zeno and had complicated theories about the divisibility of matter. Nevertheless, the idea that matter was composed of invisible and indivisible particles provided better explanations for such phenomena as condensation and evaporation, weight, and density, all explicable as a function of changing distance between particles, than did the Aristotelian elements. Once atomic theories of matter began to take hold, the defining properties of matter were no longer quality but quantity, not sensibility but divisibility and density. As atomist Kenelm Digby suggested in 1641, "In our present survay of a body, the first thing which occurreth to our sense in the perusal of it, is its *Quantity*, bulke, or magnitude," and "*Quantity* or biggnesse, is nothing else but divisibility: and that thing is bigge, by having a capacity to be divided, or (which is the same) to have partes made of it."[55]

Harriot approached divisibility as a mathematician, examining it in relation to the longstanding problems posed by Zeno's paradoxes. If a continuous line is made up of infinite points, how could that space be traversed in a finite amount of time? Harriot seems to imagine atoms as equivalent to geometric points: "in decreasing progressions we must needes understand a quantity absolutely indivisible, but multiplicable infinitely infinite till a quantitie absolutely inmultiplicable be produced, which I call universally infinite. And in increasing progressions wee must needes understand that for a last there must be a quantity inmultiplicable absolute but divisible infinitely infinite till that quantity be issued that is absolutely indivisible."[56] Harriot was unable to reconcile these speculations about commensurability and infinity with a coherent theory of material atoms.[57] However, his speculations highlight the centrality of divisibility in definitions of matter, as well as the problems posed by contemplation of insensible and counterintuitive concepts such as infinity and atoms.

The atomist theories inherited from Epicurus and Lucretius in antiquity held that the universe consisted of atoms floating in a void space which came together randomly to produce matter. The Aristotelian tradition offered a different version of a particulate theory of matter in the concept of "minima naturalia," which was developed and expanded in the Middle Ages to deal with problems of infinite divisibility and the continuation of substance in mixtures.[58] This corpuscular theory held that matter was continuous, since Aristotle had argued that a void or vacuum could not exist in nature. However, the revival of atomism in the early seventeenth century reopened the possibility of void space. If matter is composed of small, indivisible particles, what was between the particles? And what held them together?

Different thinkers treated these problems in different ways, but the longstanding Aristotelian denial of void space could no longer be taken for granted. Nicholas Hill, for instance, who was associated with the Northumberland circle, argued that "effluxus et influxus, rarefactio, condensatio, corruptio, generatio, gravitas et levitas, motus et alteratio, non immaginabilis absque vacuo [flowing out and flowing in, evaporation, condensation, corruption, generation, weight and lightness, movement and change are not imaginable without a void]."[59] As we saw in an earlier chapter, Harriot argued with Kepler about whether simultaneous reflection and refraction of light passing through various liquid media was caused by light passing through void space and bouncing off atomic particles. Harriot argues that "a dense diaphanous body, therefore, which to the sense appears to be continuous in all parts, is not actually continuous. But it has corporeal parts which resist the rays, and incorporeal parts vacua which the rays penetrate. So that refraction is nothing else than an internal reflection, and the part of the rays which are received inside, although to the sense it appears straight, is nevertheless composed of many straight line segments."[60] In a later letter he insists (in response to Kepler's view that reflection and refraction are caused by Aristotelian qualities of opacity and transparency) that "I do not understand any transparency unless caused by a vacancy" or void space between atoms.[61]

Although other thinkers argued that spirit or rays (like optical species) filled the spaces between atoms, the existence of a void in nature remained a possible implication of an atomic theory of matter.[62] As Rosalie Colie has noted, "the twinned ideas of infinity and nothing are technically dangerous; since they are so wild, at the loose edge of conceptualization and of discourse," and "psychologically destructive" as well, "threatening the familiar boundaries of human experience and of intellectual efforts to get the better of that recalcitrant experience."[63]

As Hill notes, weight, density, evaporation, and condensation were all, potentially, better explained by corpuscular theories than by Aristotelian matter theory, whether they involved a void or not. Harriot's patron Henry Percy himself argued that "the doctrine of generation and corruption unfoldeth to our understanding the method general of all atomical combinations possible in homogeneal substances," explaining "all the accidents and qualities rising from those generated substances, as hardness, softness, heaviness, lightness, tenacity, frangibility, fusibility, ductibility, sound, colour, taste, smell, etc."[64] Weight could be explained as a property of the way atoms fit together; Harriot argued that changes in weight were caused by "interposing small atoms in the *vacua* between larger ones."[65] Smell was the result of the dispersion of tiny particles in a large quantity of air, and some scientists even calculated the size of atoms based on the amount of space a particular quantity of incense could fill with odor.[66] These new speculations about the nature of materiality inform the imagery that delineates the disorderly world of *King Lear*.

In *King Lear*, Shakespeare turns again and again to problematic material phenomena like divisibility, weight, and smell to describe a world where everything seems inexplicably to be going wrong. The play offers not one coherent mental model of an orderly universe but a series of attempts to work out what the universe is made of and how its sensible properties relate to what is invisible or insensible. Viewed from this perspective, the play centrally questions whether our basic human experience of the world can, or cannot, be extended to provide an analogy that explains its workings.

In the first six lines of the play, Gloucester and Lear manifest what seems to be partial knowledge of the new ideas about matter. Both seem to know that matter is in essence divisible, but they continue to assume that immaterial phenomena manifest the same qualities and obey the same rules as the material. Both speakers assume that the kingdom is a material substance that can be divided and that this division is, in different ways, caused by or related to the "weight" of an abstraction. Gloucester notes that "now in the division of the kingdom, it appears not which of the Dukes he values most, for equalities are so weigh'd, that curiosity in neither can make choice of either's moi'ty" (1.1.2–6). The relative worth of the two dukes, and thus their shares of the divided kingdom, are imagined as having relative weight. Lear repeats the two ideas when he proclaims that

> we have divided
> In three our kingdom; and 'tis our fast intent
> To shake all cares and business from our age,
> Conferring them on younger strengths, while we
> Unburthen'd crawl toward death. (1.1.37–41)

In the one case, the worth of the two Dukes, figured as weight, is imagined as correlating with the "moiety" or share of the divided kingdom that each will receive. In the other, Lear imagines that dividing the kingdom will lessen the weight of "cares" that he carries, as if the divided portions of the material substance of the kingdom carry the weight of the abstract cares that ruling them has entailed.

It was commonplace, of course, and still is, to speak of old age, responsibility, grief, or other difficult situations as heavy burdens. This expression is an obvious mapping from a kinesthetic experience of embodiment onto abstract concepts. Shakespeare frequently does so in phrases such as "burden'd with like weight of pain" (*The Comedy of Errors*, 2.1.26) and "What passion hangs these weights upon my tongue" (*As You Like It*, 1.2.269). On the other hand, weight was also a metaphor for seriousness, gravity, and worth: "We would be resolved, before we hear him, of some things of weight" (*Henry V*, 1.2.4–5). Cicero in his *De Senectute* ("On Old Age," a common school text in the period) repeatedly refers to "our common burden of old age, " (1.2) which is "so vexatious to most old men that they declare it to be a load heavier than Aetna" (2.4).[67] In that treatise, Cato teaches that this weight of old age is, paradoxically, lighter when the aged person retains "weighty" influence (*auctoritas*) (17.81), "for old age is honoured only on the condition that it defends itself, maintains its rights, is subservient to no one, and to the last breath rules over its own domain" (11.38). Also, paradoxically, "those very things that seem light and trivial are marks of honour—the morning visit, being sought after, being made way for, having people rise at one's approach, being escorted to and from the forum, being asked for advice" (18.63).

In *King Lear*, Shakespeare does more than simply repeat these commonplace metaphors, although their relevance to Lear and his dilemma is clear. That old age is felt as weight, as well as questions about whether or not authority and responsibility add to or help to support that weight, become part of the problematic physics of the play. The play tests its characters' reliance on such metaphors and questions the assumptions about the nature of matter and the relationship between the material and immaterial that they represent.

Lear's decision to divide his kingdom and the attendant love test are integrally dependent on his and his daughters' conflicting concepts of matter, divisibility, and the correspondence between the material and the spiritual or abstract. Lear assumes that the kingdom is divisible but finite: "in three," "third," "bounds even from this line to this" (1.1.38, 80, 62). He imagines that the abstract properties of sovereignty—"rule," "interest of territory," "cares of state"—are analogous to the material substance of the kingdom, that their weight is a burden to him, and that by divesting himself of the kingdom he can also divest himself of the weight of old age (1.1.49–50). Subconsciously, he may realize Cicero's truth that it is age itself, not responsibility, that burdens the aged; even when "unburthen'd" he still imagines himself unable to walk upright, crawling toward death. More problematic—as many critics, beginning with his own fool, have realized—is his belief that he can divide "the name, and all th' addition to a king" from "sway, revenue, and execution" of the material kingdom, keeping the former while shaking off the latter.[68]

Although Marxist criticism has tended to argue that Goneril and Regan join Edgar as proponents of a new materialism, while Lear and Cordelia hold to older, spiritually informed values, their attitudes toward divisibility seem roughly opposite this received interpretation.[69] Lear asks his daughters to draw an analogy between a divided material space and the quantity of their love, equating "love us most" with "largest bounty" (1.1.51–52). Love, however, is immaterial and abstract, and it is not at all clear whether it follows the same rules of divisibility as the material of the kingdom. Goneril and Reagan understand that abstractions may not necessarily operate according to the same rules of divisibility as material substances do. Goneril likens her love for Lear to specifically immaterial phenomena: "eyesight, space, and liberty."[70] She seems to assume that love is infinitely divisible; it is "beyond" such quantitative measures as weight and value. Similarly, Reagan refers to her "deed of love," describing it as a process, not a substance (1.1.71). She represents her love as beyond what is sensible or mathematically quantifiable, what "the square of sense possesses."

Lear and Cordelia, on the other hand, seem more resolutely materialistic, assuming that love and other abstractions like "rule, " "interest," and "cares" must obey the same laws of density and divisibility as matter. Lear assumes that he can "shake off" the weight of sovereignty, along with the three parts of the kingdom, and divest himself of his cares. He takes Goneril's abstractions as commensurate with the substance of his kingdom: "forests," "champains," "meads," and "rivers" equal her "dearer than eyesight, space, and liberty."

Cordelia speaks about her love as if it were thoroughly material. It has weight when she claims that "My love's more ponderous than my tongue" and "I cannot heave my heart into my mouth" (1.1.88, 91–92). She can measure her love accurately, and it is confined within boundaries, just like the divisions of the kingdom; thus she loves "According to my bond, no more, no less," and her love is subject to finite limits: "That lord whose hand must take my plight shall carry/Half my love with him" (1.1.93, 101–2). She does the math for her sisters, based on her own assumptions about the finitude of love, and asserts that they have claimed to love Lear "all." But if love is capable of infinite division, "more than eyesight, space, and liberty" does not necessarily equal "all."[71] Cordelia's mental model of the universe is clearly more limited and constrained than that of her sisters.

Cordelia's answer to Lear's question, "What can you say to draw a third more opulent than your sisters'[?]," is literally true: there is only one third left, and she will only get that one, no matter what she says.[72] Unwilling to participate in the sisters' economy of infinitude, she insists that the kingdom and her love are materially limited. By answering, "nothing," however, she inadvertently introduces that correlative of infinity, the void, into the play. Cordelia seems to assume a kind of Aristotelian plenum—Aristotle believed that void space could not exist in the fullness of matter—and it doesn't occur to her that Lear will take her "nothing" so literally. Kent shares her perspective when he insists, "Nor are those empty-hearted whose low sounds/Reverb no hollowness" (1.1.153–54), attempting to convince Lear that her "nothing" does not indicate empty space. But Lear, believing in a correspondence and commensurability between matter and the insensible, perhaps influenced by the concept of infinity that Goneril and Regan have introduced, seems to take "nothing" more literally than Cordelia intends it. To say "nothing" potentially indicates the presence of void space behind it. Cordelia herself, Lear imagines, is emptied out, and he calls her a "seeming substance . . ./And nothing more" (1.1.198, 200). His answer, "nothing will come of nothing," thus, I think, echoes the atomists' sense that it is necessary for "nothing" to exist between and around the indivisible particles of divisible matter. "Nothing" for him opens up an ontological and epistemological chasm into which he falls, headlong, dragging everyone with him.

Lear remains preoccupied by problems of divisibility and materiality throughout the play. His warnings to Kent, for instance, later in this scene reflect his growing sense that if matter can be divided from itself, then it can also be divided from the immaterial and spiritual abstractions that have ordered it. He commands Kent to "come not between the dragon and his wrath" (1.1.121) nor "to come betwixt our sentence and our power" (1.1.170). He fixes on his one hundred knights as a material correlative of the abstract authority he believes he has retained, so his obsession with refusing to divide their number is not surprising, nor is his anger that "She hath abated me of half my train" (2.4.159). He insists that "thy half o' th' kingdom . . . which I thee endow'd" (2.4.180–81) ought, by some sort of reverse proportionality, to keep the sisters from forcing him to halve his retainers.

The fool tries on various attitudes toward divisibility and substance. His advice, "have more than thou showest, speak less than thou knowest," suggests a kind of conservation of matter, advising Lear to make sure that any division of resources goes in his favor. The logic of "more" and "less," rather than "all" or "none," identifies a middle range between infinity and void. Lear, however, refuses to accept that self-discipline might compensate for the terrifying new properties that matter seems to have taken on, stretching into infinity on the one hand, falling away to nothing on the other. This advice seems like "nothing" to him and causes him to reiterate the idea that "nothing can be made out of nothing" (1.4.133). The fool then takes Lear's own logic to an extreme, insisting that Lear's abstract authority is indivisible from his material rule and insisting also that the site of a division is void space. Zero, the cipher, stands for the void when it is decoupled from a limiting figure: "thou hast par'd they wit o' both sides and left nothing i' the middle" (1.4.187–88); "now thou art an O without a figure" (1.4.192). As Lear had earlier accused Cordelia of being "seeming substance," the fool now identifies Lear as "Lear's shadow."[73]

The dangerous possibility of "division" also haunts the play in the form of a projected civil war between Albany and Cornwall, which, famously, never happens. Kent comments that "There is division/(Although as yet the face of it is cover'd/with mutual cunning) 'twixt Albany and Cornwall" (3.1.19–20), and Gloucester also notes "There is division between the Dukes" (3.3.7–8).[74] Gloucester's attribution of occult causes for the proliferation of divisions in the play looks back to the kind of conjunction of matter and spirit that the play in other respects questions. It is significant that "division" is the central symptom of the troubling chaos that Gloucester believes to be caused by "these late eclipses": "brothers divide," "the bond is crack'd twixt son and father" (1.2.107, 108). Most critics agree that the play dismisses Gloucester's appeal to astrology as a causal and explanatory force, and I would argue that the proliferation of divisions (even potential divisions that don't, in fact, occur) emerges from an atomistic model of matter. In Kenelm Digby's words, "*Quantity* or bignesse, is nothing else but divisibility; and . . . a thing is bigge, by having a capacity to be divided, or (which is the same) to have partes made of it" (9). The play recognizes the social implications of division and discord, and, of course, there was a longstanding tradition that dividing a kingdom would cause further rifts. But this play's divisions are also based in a deeper ontological rift between matter and spirit and within matter by its very nature. Gloucester's belief that the stars and planets give out rays that limit and control elemental matter is shown to be outmoded.

As Henry Turner has argued, Lear's madness takes place not (as the editorial tradition falsely suggests) on a "heath" but in a kind of void space. Lear "wanders 'without' into a breach in fictional space to flail in the *potentia* that surrounds him, a point somewhere *between* a coherent location and the open stage."[75] Turner associates this space with "the quasi-scientific space of the *vacuum* or *void*" (177). Lear ends up in this space because he has suddenly become aware of it. His voicing of Harriot's "nothing will come of nothing" opens up the vacuum that exists within the fabric of divisible substance, all around it, and between the sensible experience of the material world and its meaning. When Lear asks the thunder to "strike flat the thick rotundity o' th' world, / Crack nature's moulds, all germains spill at once / That makes ingrateful man" (3.2.6–9) he calls up an absolute void. Nature's molds are the Aristotelian forms that, in some theories, give ordered shape to matter; the "germains" are potentially the life-giving seeds of Neoplatonic tradition or the lifeless atoms of Harriot's theory or, perhaps, both at once.[76] Either way, destruction of seed and mold would leave a void, from which nothing could ever be made again.

In this context, Lear's idea about the division of wealth, that the wealthy should "shake the superflux" to the poor, seems inadequate. There is certainly no superflux in a void. After his experience of nothingness, however, Lear loses his preoccupation with divisibility and stops trying to understand how matter and the immaterial are joined. His misogynist ramblings, for instance, offer up a complicated supernatural scheme to describe what Shakespeare in the sonnets and elsewhere calls the "nothing" between women's legs.[77] For Lear, the sulphurous pit may be less frightening than contemplation of another void. Instead of attempting to control matter, Lear comes to rely on his own feelings as the one clearly present phenomenon; his and Cordelia's tears and his own sufferings become the only tangible evidence of existence. In this respect, Lear joins other characters who come to rely on weight and smell, rather than divisibility and sight, to make their way in the world. If embodied experience does not any longer yield the truth about the invisible causes of the material world or about the immaterial ideas that help us interpret it, it may be best to dispense with understanding and embrace sensations of embodiment in and of themselves.

Weight, the universal experience of gravitation, becomes a crucial concept for Edgar and Gloucester as well as for Lear and represents an experience of the material that might seem less readily divisible. We recall Digby's arguments that "dense" or heavy bodies are less prone to division because their atoms "are more close and compacted, then others have that are more rare and subtle" (16). Lear and Cordelia, however, attempt to endow abstract qualities—care and love—with weight, in order to make them substantial and finitely divisible. In these efforts, they resemble Dee's speculations about "Statike" in his preface to Euclid, as well as Spenser's Giant. On the other hand, Lear errs in believing that he can separate the burden of cares from the weight of old age itself. It was common to imagine the weight of old age as positive when it brought with it "gravity" and wisdom; however, Lear, as other characters repeatedly note, lacks this ballast: "he hath ever but slenderly known himself" (1.1.293–94); "As you are old and reverend, should be wise" (1.4.240); "Thou shouldst not have been old till thou hadst been wise" (1.4.44). Because of this lack, Lear seems to imagine weight as almost always negative, although the consequence of this imagining turns out to be worse—the weightlessness of the void.

Edgar and Gloucester begin by associating weight with suffering and burden but come to see it as a necessary corollary of existence. Finally, you know you are alive because you feel your own weight and are subject to gravity. Edgar views his own and Lear's sufferings as weight they bear: "we see our betters bearing our woes." He imagines that this burden is eased when other people seem to share it: "How light and portable my pain seems now" (3.6.102, 108). Gloucester, similarly, describes his planned suicide as intended to "shake patiently my great affliction off" since he is unable to "bear it longer" (4.6.36, 37).

The feigned jump off Dover cliff is, in part, an exercise in the experience of weight and gravity. Henry Turner has suggested that the lines in which Edgar conjures up for his father the illusion of climbing the cliff are based on new concepts of space, "substituting a *verbal* description for the geometrical forms and mathematically-derived proportions typical of perspective painting, and then, additionally . . . turning this illusion into a second-party narration for a blind man who, after all, *cannot see anything.*"[78] As such, the scene emphasizes the potential gap between visual experience of the material world and the reality that lies beneath, an experience that the audience is made to share in this scene. The force of gravity here becomes a more accurate indication of the true situation. When Edgar insists, "You do climb up it now. Look how we labor," Gloucester replies, "methinks the ground is even" (4.6.2–3). Edgar insists again, "horrible steep," and suggests that Gloucester's other senses "grow imperfect by your eyes' anguish" (4.6.3, 5), trying to create an illusion that will distract Gloucester from his accurate sensation of gravity.

After Gloucester has experienced his "jump" off the cliff, Edgar continues to "trifle" with his father's sense of weight: "Had'st thou been aught but goss'mer, feathers, air / (So many fathom down precipitating) / Thou'dst shiver'd like an egg" (4.6.48–50). Yet, Gloucester is still in one piece and still has "heavy substance" (4.6.52). When Edgar proclaims, "thy life's a miracle," that miracle seems based on the concept of a paradoxically weightless matter, "goss'mer, feathers, air," and yet imbued with "heavy substance." The conclusion that he wants Gloucester to reach is, not surprisingly, that he must withstand the weight of existence, "bear / Affliction" until it is his appointed time to die.

As Bruce Bartlett has noted, Shakespeare sometimes uses the word "Waight" to pun on "weight" and "wait."[79] When, in the final lines of the play, Edgar says, "The weight of this sad time we must obey," he refers to the heavy burden of suffering and responsibility ("the oldest hath borne most") and also to the "sad time" that must be endured on Earth. Bartlett is right, I think, to draw a connection between this "weight" and Edgar's lessons to Gloucester: "men must endure / Their going hence even as their coming hither. / The ripeness is all" (5.2.10–11). Gloucester is able to complain that "I stand up, and have ingenious feeling / Of my huge sorrows" (4.6.280–81), and that feeling is in part his sense of weight and gravity, imagined as a tangible experience of sorrow—signs that he is still alive. While this sense of weight seems more accurate than sight in conveying the experience of life, it is not foolproof, as Edgar himself has shown. If Edgar has attempted to demonstrate that the weight of mortality is not always deadly, he has done so only by "trifling" with Gloucester's sense of gravity.

When Lear enters "*with Cordelia in his arms*" (5.3.257ff.), he, like Edgar, contrasts her heavy body—"dead as earth," the densest element—with the "feather" and "breath" that might indicate life. Lear seeks visual confirmation that she lives: "if that her breath will mist or stain the stone" (5.3.263) or if the "feather stirs, she lives!" Or, again, "look at her lips." These light signs of life are imaginary and have no weight to counter the body, "dead as earth," or to counter Lear's experience of that burden. The densest matter seems here to be the most substantial, although Lear dies yearning for visible proof of something beyond it.

If visible assessment of the material world is faulty and a sense of weight can be relative, the play also suggests that phenomena associated with the other material extreme, rarity, can provide a more accurate perception of the nature of reality. Theorists of corpuscularism were interested in smell as a manifestation of matter at its least dense. Seventeenth-century scientists studied smell as an instance of "transport phenomena," depending on the existence of tiny particles to explain how "material substances [could] appear or disappear unnoticed." Smell was explained as "an efflux of corporeal particles."[80] The play repeatedly offers smell as an alternative to vision or experience of weight as a means of gaining more accurate knowledge about the material world.

Robert Heilman traces images of smell through the play, arguing that they represent the "general corruption" of the world.[81] Certainly, many of these images do suggest the stench of corruption: "there's the sulphurous pit, burning, scalding, stench, consumption" (4.6.129–30). But in the context of the play's interest in human experience of materiality, smell takes on additional significance, as a way of representing a direct but uncontrolled experience of the fabric of existence. Eve Sweetser has charted the role of different metaphors of sense perception, as they are used to represent mental states in English and other Indo-European languages. As part of the cognitive system through which we understand abstract phenomena in terms of physical experience, we persistently imagine the mind in terms of the body. Thus, mental perception is almost always conceived in terms of sense perception. Sweetser shows that there is extensive correlation across languages among the specific mental qualities that are conceived in terms of specific physical senses. Metaphors of vision pertain to intellection, "and much of the detailed vocabulary of our visual domain can be used to structure the description of our intellectual processes" (I see, you've made that clear, he's in the dark, she's brilliant).[82] Metaphors of hearing, on the other hand, convey "receptivity," "heedfulness," and "obedience," while touch is used to suggest control, understanding, and emotion, and taste, "personal preference."[83] Smell is related to touch, but Sweetser concludes that "smell has few abstract or mental connotations, although bad smell is used in English to indicate bad character or dislikeable mental characteristics."[84]

Smell becomes important in *King Lear* for just this reason: it is not a part of the usual human interface between concrete and abstract experience, between what we can physically experience and what we imagine our minds are doing. Smell seems less controllable than vision or hearing (it can't be blocked out or focused) and less personal than taste. In the corpuscular materialism of the seventeenth century, however, smell offered direct experience of the tiny particles that made up all matter; spread widely through the air, they were actually inhaled into the body. Galenic medicine had long identified "corrupt air," detectable by its putrid smell, as a cause of disease, and early modern medical writers like Fracastoro and Paracelsus suggested that "seeds" of contagious disease were carried in the air.[85] Smell in *Lear* thus conveys the sense that is left to rely on when all of the usual ways of extending the senses metaphorically to understand what we can't directly perceive have failed. It also offers, perhaps, a distant possibility of direct, rather than metaphorical, participation in the material world.

In the play, smell is sometimes explicitly imagined as substituting for other failed senses. When Gloucester has been blinded, Regan suggests that he "smell / His way to Dover" (3.7.93). Smell is also offered as a way to perceive flaws or faults that are hidden from sight, as when Lear claims that he was fooled by flatterers until "when the thunder would not peace at my bidding, there I found 'em, there I smelt 'em out" (4.6.102–3). The Fool, on the other hand, casts doubts on the human capacity to detect by smell what can't be seen when he tells Kent that "all that follow their noses are led by their eyes but blind men, and there's not a nose among twenty but can smell him that's stinking" (2.3.69–71). He argues that the nose is in the middle of the face "to keep one's eyes on either side's nose, that what a man cannot smell out, he may spy into" (1.5.22–23).

Lear's misogynist fantasy links smell to smoke, another phenomenon explained more easily in terms of tiny particles, when scientists compared "the smoky wick of an extinguished candle, which was not even the size of a pea, and the huge volume of air that was filled with its smoke."[86] Lear's description of women's bodies below the waist, "the sulphurous pit, burning, scalding, / Stench, consumption"(4.6.127–28), links the stench literally with burning, not decay. Smell here represents the particulate effluvium of materiality and mortality, as it also does when Lear insists that his hand "smells of mortality" that he must try to wipe off (4.6.133). When he sums up the human condition by noting that "the first time that we smell the air / We wawl and cry" (4.6.179–80), he is not necessarily suggesting that all existence bears the stench of corruption (as Heilman argued) but that the air carries the invisible residue of the material world. We cry because of our birth into materiality and mortality.

King Lear, I have been arguing, represents a world in which the "fit" between perception and reality has been questioned, and, as a result, human beings feel radically dispossessed. The material world has lost its solidity and exhibits alarming new properties; it may be infinitely divisible and riddled with void space. Matter may manifest itself as burdensome weight or as the mere atomic trace of a bad smell. And when the basic spatial experiences of embodiment are disjoined from human understanding of what cannot be seen or felt, the ontological and epistemological crisis that is *King Lear* is a result.

It's worth asking how this basic dislocation of materiality relates to the other kinds of dispossession and chaos that critics have traditionally identified in the play. Certainly, it is easily related to most of them. For instance, when Richard Halpern argues that "*King Lear*, I think, is largely about the divorce between the signs and the material realities of royal power," he addresses a social disjunction that mirrors the epistemological rift I have been describing.[87] However, it remains to ask whether one of these "causes" is prior to the other. Does a breakdown in a mental model of the physical world precede and cause an ideological shift, such as the one Halpern describes? Or, as we have been taught to think after Foucault, does an ideological shift (from feudalism to capitalism, for instance) entail an accompanying shift in the generally accepted model of the physical world?

It is perhaps tempting to assert that the image schemas through which we conceive and think about the world, arising as they do from preverbal experiences of spatial embodiment, are prior to social formations and shape them. On the other hand, as soon as we begin to put these experiences into words, they are shaped by the ideological forces that structure culture and language. Priority is thus probably indeterminate, but it is important to realize that all ideological structures grow out of and are intertwined with basic models of the universe and how it works. Critics who read *Lear* in light of social change often produce readings that are haunted by its engagement with the spatial experience of matter. Steven Greenblatt, after all, persistently describes the relation of the play to religious belief in terms of hollowness or emptiness. He points to "the evacuation of the divine presence from religious mystery, leaving only vivid but empty ceremonies," as well as "rituals and beliefs that are no longer efficacious, that have been *emptied out*." Furthermore, "the official position is *emptied out*, even as it is loyally confirmed," and theater "evacuates everything it represents. By doing so the theater makes for itself the hollow round space within which it survives."[88] Greenblatt, of course, picks up on the play's own obsession with empty space, and it is a strength of his reading that he phrases his discussion of the incipient atheism of the play in its own terms; he shares Shakespeare's mental model, and his criticism participates in it. Richard Halpern's Marxist reading of *King Lear* is similarly in touch with the physicality of the material, using repeated images of weight and quantity, almost apologetically noting that "the play (if I may be permitted another physics metaphor) calculates something like a 'sum over histories,' assuming that *every* possible historical path is taken, and averaging out the results."[89] Again, I would argue that Halpern picks up on and uses the play's mental models and that his reading of its political and economic investments is strengthened by relating them, even unconsciously, to the material substratum of imagery. The language of poetry, especially that of a poet like Shakespeare who is in the business of conjuring up a world largely out of language, will necessarily be shaped by the poet's interactions with the physical world and his or her models for how it works. Imagery and metaphor convey these models so that the models are woven into the fabric of poetry and provide a structure or frame that supports plot, characterization, and theme. The history of science is invaluable in helping us parse the unfamiliar shapes that such models could take in other times, and it is important to remember that, in most cases, scientific knowledge does not shape a poem directly but does so for the most part only insofar as it informs the writer's basic mental models. Good readers of a text may sometimes intuit the shaping presence of these models, but a cognitive history of ideas can greatly enrich our experience of a text like *King Lear*.

CHAPTER 6

Matter and Power

Critics since Marjorie Hope Nicolson and E. M. W. Tillyard have emphasized the social implications of the hierarchically ordered Aristotelian cosmos, calling attention to passages like Ulysses' speech in *Troilus and Cressida*, in which "the sun, the king, primogeniture hang together; the war of the planets is echoed by the war of the elements and by civil war on earth."[1] In this book so far, I have been suggesting that the breakdown of this world picture and the concomitant divorce of intuitive experience of nature from scientific theory carried epistemological and emotional consequences that can be seen as separate from (even if they were in many ways related to) anxiety about social disorder. People were upset about a loss of direct access to truth about the natural world regardless of whether this change also portended social chaos. However, as we saw in chapter 5, basic ideas about the nature of numbers and matter inevitably have implications for how someone like William Shakespeare thought and wrote about social relationships and political structures.

In this chapter I look more explicitly at the different ways in which Christopher Marlowe and Shakespeare imagine the projection of power and colonial domination in terms of changing ideas about the cosmos and the elements. Both writers reflect new attitudes toward meteorology that emerged at the end of the sixteenth century in England. Marlowe uses new work in cartography and astronomy to find a language adequate to convey Tamburlaine's immense ambition and will to conquest. Shakespeare's *Antony and Cleopatra*, on the other hand, represents the waning elemental physics as stubbornly resistant to conquest by a new hard-edged physics of surfaces. Both writers are less directly concerned about the hierarchical arrangement of the cosmos and more attuned to the contemporary problems in matter theory that I have been tracing throughout this book. Marlowe seems to have been more interested in the possibilities offered by a universe newly secularized and open to change, whereas Shakespeare seems to have been disturbed by the implications of losing an understanding of the world that seemed more accessible to and intertwined with the human senses.

Caroline Spurgeon was correct in her observation that Marlowe's repertoire of images was upward looking, with images of "the sun, moon, planets, and heavens" outnumbering all others, evidencing "an imaginative preoccupation with the dazzling heights and vast spaces of the universe."[2] She identifies Shakespeare's imagery as essentially Earth based, with nature, especially weather, plants, and animals, predominating (13). Marjorie Hope Nicolson used Spurgeon's study of Marlovian imagery to conclude that had Marlowe lived long enough to become aware of the new science, he, unlike Donne, would have been excited about it.[3] However, Nicolson laments that Marlowe's universe remains thoroughly Ptolemaic and Aristotelian.

As we saw in the first two chapters, Nicolson was incorrect to claim that Marlowe "died too early" to know about the "new astronomy." Robert Recorde alluded to Copernicus's heliocentric model in 1556, and Thomas Digges published a translation of part of the *De revolutionibus* in 1576. We have seen evidence that educated Londoners in the 1580s and 1590s, people like Gabriel Harvey and Thomas Harriot with whom Marlowe was acquainted, knew about the Copernican model of the universe.[4] Critics have noted Marlowe's use of cutting-edge cartography in *Tamburlaine*, his most "scientific" play.[5] But scholars have been less clear on what is new in his meteorology and astronomy.[6]

It is true that descriptions of the universe in *Tamburlaine* and in Marlowe's other plays are resolutely Ptolemaic. But, as we have seen, ability to imagine the physical structure of the universe in Copernican terms is not, in fact, a reliable criterion for determining whether or not the "new astronomy" influenced the person doing the imagining. For one thing, the ability to picture a heliocentric universe lagged behind its acceptance in theory. In sixteenth-century England, knowledge of and even belief in the correctness of Copernicus's mathematical account of the movement of the heavenly bodies did not necessarily mean acceptance of his theory as a physical model of the universe. Those mathematicians who were in a position to understand the superiority of Copernicus's calculations—John Dee, Robert Recorde, Thomas Digges, Thomas Harriot, and others—struggled to come to terms with the fact that the appearance of the heavens as observed from Earth supported the Ptolemaic model.[7]

We can identify three areas of innovation and speculation that provide a context for the moral universe of Marlowe's *Tamburlaine*. First, as several scholars have noted, advances in cartography and the publication of maps that allowed people to visualize an expanding world gave Marlowe a sense of imaginative control over large geographical spaces, which provides a setting for Tamburlaine's imperial ambitions and the basis for Marlowe's verbal technology for projecting an illusion of vast space on the platform stage.[8] Second, English Protestants invested Aristotelian meteorology with a newly intensified naturalism, arguing that because divine intervention in nature no longer occurred, meteorological events like comets and earthquakes were not to be read as signs of God's moral approbation or disapprobation but simply as natural events caused by atmospheric conditions. Marlowe's meteorological images in Tamburlaine reflect this new sense of a world abandoned by God to human devices and ambitions. Finally, new technologies for determining the distance and location of astronomical phenomena meant that the supernova of 1572 could, for the first time, be definitely identified as a new event taking place above the moon, a realm that was supposed to be immune to change and that therefore guaranteed the stability of the universe as a whole. While part 1 of *Tamburlaine* confines Tamburlaine's projected power to the area of the heavens located below the moon, part 2 uses images that repeatedly imagine Tamburlaine's power as projected above the moon and able to threaten the stability of the highest reaches of the heavens. The play mirrors the ambivalence with which educated Londoners greeted new ideas about the universe; Marlowe, like many of his contemporaries, seems to have experienced both excitement and pessimism in response to them.

Critics have long been aware that Marlowe made use of Abraham Ortelius's recently published world atlas, the *Theatrum Orbis Terrarum* (1570), as a source for the geography of Tamburlaine's empire.[9] John Gillies has argued that Ortelius's atlas was "new in the sense of adding unknown continents" and also in the sense that it represented space in a way that was newly "indifferent to religious symbolism" (38). The atlas inspired Marlowe to express his "hero's ambition and potency" in terms of "an imagery of sheer geographic extent" (37). Henry Turner argues that Marlowe was also inspired by maps to fundamentally reimagine and expand the resources of the platform stage, using "geometrical projection as a kind of poetic projection that is somehow necessary to stage performance."[10]

If Ortelius's maps helped Marlowe give imaginative shape to Tamburlaine's imperial ambitions and also inspired a technology for realizing his own ambitions for the Elizabethan stage, contemporary meteorological writing helped him create a three-dimensional space for Tamburlaine's feats, making it possible to imagine an upward trajectory linked to human elemental nature and uninhibited by the kind of moral judgments previously thought to be expressed by God through the natural world.[11] In 1 *Tamburlaine* 1.2.48–51, Tamburlaine uses terms familiar from discussions of the 1580 earthquake when he boasts to Zenocrate that his army will be "So great an host,/As with their waight shall make the mountains quake,/Even as when windy exhalations,/Fighting for passage tilt within the earth."[12] We can recall that Aristotelian meteorology was based on the idea that many atmospheric phenomena were caused by dry "exhalations" given off by earth or moist "vapors" given off by water which rose up (pulled by the heat of the sun) through the three regions of the air (a warm layer close to the earth, a cold layer in the middle, and a hot layer close to the sphere of fire).[13] Dry exhalations trapped within the hollows of the earth caused earthquakes as they tried to escape.[14]

In classical antiquity, unusual meteorological events such as comets and earthquakes were believed to foretell disasters. Christian writers adapted this idea to claim that comets and earthquakes represented a warning from God about impending events. In the introduction and in chapters 2 and 3, I traced a range of responses to meteorological events in late sixteenth-century England, ranging from Arthur Golding's insistence that they represent " infallible forewarnings of Gods sore displeasure for sinne, & of his just plagues for the same, where amendement of lyfe ensueth not" (B1r), to Gabriel Harvey's conclusion that "an Earthquake might as well be supposed a Naturall Motion of the Earth, as a preternatural, or supernaturall ominous worke of God" (59), to arguments by Protestant writers like William Fulke that God's miraculous interventions in nature had ceased and therefore the supernatural interpretation of earthquakes, comets, and blazing stars was doubtful.[15]

We can now appreciate the audacity of Tamburlaine's claim that the weight of his vast army will cause earthquakes. Rather than taking the older view that earthquakes are supernatural events representing a warning from God that wars or other disruptions are about to occur, Tamburlaine claims that his army itself is heavy enough to shake the Earth. He eschews a supernatural interpretation; he, not God, has the power to shake the Earth, and he is able to do so without employing the natural mechanism of trapped exhalations.

Tamburlaine and the other characters in the play frequently use meteorological metaphors and analogies to measure and explain human ambition. His famous assertion in 1 *Tamburlaine* that "Nature that fram'd us of foure Elements,/ Warring within our breasts for regiment,/Doth teach us all to have aspyring minds" (2.7.18–20) is, as Gillies has argued, an unorthodox claim in terms of Galenic medicine, where a state of health demanded that the humors be in balance.[16] In meteorological theory, however, the four elements (which corresponded to the four humors) did vie with each other, and air and fire did "aspire" to rise above heavier earth and water. Theridamas constrasts Cosroe with Tamburlaine in elemental terms: "For he is grosse and like the massie earth, / That mooves not upwards, nor by princely deeds / Doth mean to soare above the highest sort" (2.7.31–33). Tamburlaine's warring elements create a kind of internal earthquake as his airy spirits attempt to break free from his grosser parts.[17]

In an extended passage in 2 *Tamburlaine,* he provides his own naturalistic reworking of the premonitory power of meteors, replacing supernatural causation with his own agency. His burning of the town where Zenocrate has died will send flames "to the highest regions of the aire, / And kindle heaps of exhalations, / That being fiery meteors, may presage, / Death and destruction to th' inhabitants!" (3.2.2–5). His destruction of the town is imagined as the direct cause of meteors, which thus stand as signs of the destruction that caused them. He goes on to imagine that his triumph over the town will cause

> a blazing star,
> That may endure till heaven be dissolv'd,
> Fed with the fresh supply of earthly dregs,
> Threatening a dearth and famine to this land,
> Flying dragons, lightening, fearfull thunderclaps,
> Sindge these fair plaines. (3.2.6–11)

Here, the exhalations sent up by his destructive activities will create a range of meteorological effects that, instead of foretelling upheaval and destruction, will themselves damage their lands. Since this passage appears in part 2, there are hints that these effects extend beyond the atmosphere, creating a "blazing star" like that discussed by Thomas Digges and Francis Shakelton but attributed here to Tamburlaine instead of God.

In part 1, however, Tamburlaine's aspirations are bounded by the limits of the Aristotelian cosmos; they do not ascend by natural means above the lunar sphere. He describes aspiration to Theridamas in entirely orthodox terms: they will be united "Untill our bodies turn to Elements: / And both our soules aspire celestiall thrones" (1.2.236–37). Only after death is the soul imagined as ascending into the heavens. In this same scene he suggests that the heavenly spheres are stable and unchanging: "Sooner shall the Sun fall from his Spheare, / Than *Tamburlaine* be slaine or overcome" (1.2.176–77).

Only once in part 1 does Tamburlaine seem to claim that he can project his transformative ambitions into the realm of the fixed stars, but his claim turns out to be based on an optical illusion, like those described by Fulke to explain away strange visions in the sky: "And with our Sun-bright armor as we march, / Weel chase the Stars from heaven, and dim their eies/That stand and muse at our admyred armes" (2.3.22–24). Although it will appear as if the stars have been dislodged, this illusion is caused by the dazzling brilliance of sunlight reflected in the armor of his soldiers.

As we have seen, Marlowe's contemporaries reacted in different ways to the news that the new star of 1572 was located above the moon. Thomas Digges believed that it was a miracle, analogous to the star created by God to herald the birth of Christ. Others, like Shakelton in *A Blazyng Starre* (London, 1580), after acknowledging the natural causes of most comets and meteors, cited Digges to argue that the star had been sent by God to "stirre up by all the sorte of us (beeyng most wretched and miserable sinners) to be mindfull of the Judgment daie," a final miracle heralding the end of the world (Biv).

For Marlowe, the removal of this limit to change and mortality provides a setting for Tamburlaine's more extreme claims in part 2. In part 1, as we have seen, Tamburlaine aligns his extended power with geographical expansion across the surface of the Earth and up into the atmosphere below the moon. In part 2, however, after the death of Zenocrate, he, like Spenser's Mutabilitie, transgresses the lunar boundary, repeatedly imagining himself as able to change or destroy the realm of the fixed stars. These imaginings signal the excessive and destructive nature of Tamburlaine's aspirations, and it is this imagery, perhaps more than his burning of the Koran, that suggests he may have gone too far.

Immediately after Zenocrate dies, Tamburlaine claims that in his grief he will

> Raise Cavelieros higher than the cloudes,
> And with the cannon breake the frame of heaven,
> Batter the shining pallace of the Sun,
> And shiver all the starry firmament:
> For amorous *Jove* hath snatcht my love from hence. (2.4.104–8)

There is blasphemy here, of course, in his attack on Jove. But he spends three lines describing his assault on the "frame of heaven," "palace of the sun," and "starry firmament," all of which were newly vulnerable after 1572.

Part 2 contains many such imagined transgressions of the heavens. Tamburlaine says he will "with the flames that beat against the clowdes,/Incense the heavens, and make the starres to melt" (4.1.195). He has the power to extend meteorological phenomena up into the heavens, making "Meteors, that like armed men/Are seene to march around the towers of heaven,/Run tilting round about the firmament,/And break their burning Lances in the air" (4.1.203–6).[18] As he dies, using a map to lament the geographic areas that he will never conquer,

Tamburlaine leaves his sons with a warning that echoes the ambivalence of his imagined transgressions of the highest heavens. He urges them to "be warned" by the example of Phaethon, whose attempt to drive the chariot of the sun was ended by Jupiter. His warning is not that they avoid a similar transgression but rather that they take care to guide the chariot as strongly and as competently as their father would. In writing *Tamburlaine*, Marlowe's imagination was clearly imbued with the cosmological controversies of his day, and they contribute to his exhilarating but uneasy exploration of a universe newly devoid of traditional judgments and limits.

In subsequent plays Marlowe observes the lunar boundary, turning his attention to other, human-scale forms of transgression. Heninger notes that Marlowe uses more images from meteorology in *Tamburlaine*, "where excess is the rule," but because of "decorum," "marvelous meteors are all but absent from his other work."[19] Faustus's dragon chariot allows him to see but apparently not to visit the highest reaches of the heavens.

> Learned Faustus,
> To know the secrets of astronomy,
> Graven in the book of Jove's high firmament,
> Did mount himself to scale Olympus's top,
> Being seated in a chariot burning bright,
> Drawn by the strength of yoky dragon's necks,
> He now is gone to prove cosmography.[20]

Mephistopheles refuses Faustus's command to "make the moon drop from her sphere," protesting that he can only perform what Lucifer approves (1.3.283, 287). He answers Faustus's cosmological questions with orthodox Ptolemaic theory, and Faustus realizes that the answers are obvious: "these slender trifles Wagner can decide" (2.3.677). Despite Faustus's early claims that magic will provide a source of knowledge about the world to exceed all established university disciplines, the play does not reflect the same awareness of contemporary ideas about the nature of the cosmos that *Tamburlaine* does.

The expansive geography in *Antony and Cleopatra* has been compared with that of *Tamburlaine*, although Shakespeare's lists of exotic place names seem to have been inspired by similar lists in Plutarch's *Life of Antony* rather than the maps in Ortelius's atlas.[21] Although Shakespeare may not have been on the cutting edge of cosmography, he does seem to have been well versed in meteorology. As we saw in chapter 5, S. K. Heninger has established Shakespeare's knowledge of Aristotelian ideas about the elements and notes that "his works abound in the technical terminology associated with" meteorology.[22] Like Marlowe, Shakespeare seems to have been aware of ongoing debate about the natural or supernatural causation of meteoric events and to have tended toward a naturalistic reading of them. Although Heninger argues that "prodigious meteors occur prominently in Shakespeare's works, especially in the history plays where they foretell future political events," it is more accurate to say that the history plays abound in characters who *claim* that they can interpret meteors. Welsh characters are especially prone to such claims, as when the Welsh captain in *Richard II* claims to know that Richard II is already dead in Act 2 because "meteors fright the fixed stars of heaven" (2.4.9).[23] In *King John*, Pandulpho comments on the fact that discontented subjects will "pluck away the natural cause" of a "natural exhalation in the sky" and instead "call them meteors, prodigies, and signs" (3.4.153–57). Rebecca Totaro has argued that curses in the first tetralogy are linked to naturalistic physiological and meteorological phenomena.[24] Shakespeare thought in naturalistic, elemental terms and seems to have seen the world as operating through natural exhalations and vapors.

THE ELEMENTS AND POWER IN SHAKESPEARE

Shakespeare's *Antony and Cleopatra* probably dates from around 1606–8, fairly soon after his meditations on matter, nothingness, and void space in *King Lear*. This play sometimes resembles *Tamburlaine* in its conjuring up of vast tracts of geographic space and its interest in human ambition and conquest. Unlike Marlowe, though, Shakespeare remains rooted in terrestrial physics, contrasting the traditional system of the elements and exhalations with a new vision of the world that radically separates the human body from a natural world that it surveys and controls. Unlike *King Lear*, which explicitly engages with new ideas like an atomic theory of matter riddled with void space, *Antony and Cleopatra* remains vague about the theory of matter that subtends the Romans' new world of surfaces. This play is backward looking and nostalgic, focused on what it means to lose a grounding in the elements rather than concerned with what specific new concepts might be replacing them.

Critics over the years have found many ways to read the binary division of Shakespeare's *Antony and Cleopatra* between the poles of Rome and Egypt.[25] Recently, postcolonial theory has informed readings that emphasize the "Otherness" of Egypt: as John Gillies has argued, the "'orientalism' of Cleopatra's court—with its luxury, decadence, splendour, sensuality, appetite, effeminacy, and eunuchs—seems a systematic inversion of the legendary Roman values of temperance, manliness, courage, and *pietas*."[26] However, as these critics usually acknowledge, the contrast between the two blurs upon closer inspection, since, as Gillies again puts it, it is "only from the vantage point of Egypt that Rome actually seems 'Roman.'"[27]

I want to approach the differences between Rome and Egypt in Shakespeare's play as, in large part, cognitive differences, based in Shakespeare's imaginative engagement with changing theories of the relationship between human sense perception and scientific truth. Rome and Egypt seem to be the sites of different perceptual styles, which are in turn based upon different beliefs about the nature of the material world. The cognitive orientations of Rome and Egypt have different epistemological underpinnings and political implications. Romans in the play name their environment the "world" and perceive and understand it primarily in visual terms. Their "world" is composed largely of hard, opaque, human-fashioned materials, and its surface is divided into almost obsessively named—and conquered—cities and nations. Caesar refers to the reaction of the "round world" to Antony's death (5.1.15), and a temporarily Romanized Antony warns Octavia that "the world and my great office will sometimes / Divide me from your bosom" (2.3.1–2).[28] Egyptians, on the other hand, inhabit the "earth," in which they imagine themselves to be immersed and which they perceive and understand through all of the senses. The "earth" is yielding, encompassing, generative, and resistant to human division and mastery: a defeated Antony asks that Caesar let him "breathe between the heavens and earth / A private man in Athens" (3.11.14–15), and after Antony's death, Cleopatra cries "the crown o'th'earth doth melt" (4.16.65).[29]

As William Cunningham points out in his *Cosmographicall Glasse* (1559), early modern English used "worlde" to denote the object of cosmography, the study of the Earth and the heavens. "Th'earth" on the other hand, was for Cunningham the object of geography, which studied "Hylles, Montayns, Seas, fluddes, and such other notable thinges, as are in it conteined."[30] In Shakespeare's play, Egyptian understanding of their relation to the Earth is partly based on the Aristotelian system of elements and humors that was, by 1606, as we have seen, at the beginning of the end of its dominance. Romans, on the other hand, seem to have left behind that system and its porous interrelationships between subject and nature, replacing it with a subjectivity separated from and overlooking the natural world and imagining itself as able to control it. These differing systems of thought and perception result in different versions of nation and empire. The Roman "world" seems to be reaching toward something like what Shankar Raman has termed "colonialist space" and toward the rational subject who can exploit it.[31]

Egyptian earthiness suggests the intractability and inscrutability of nature in the face of the human will to power.

The attractiveness of Egypt and unattractiveness of Rome have troubled many critics, and Shakespeare's relatively positive representation of Egypt has sometimes been read as nostalgia for an heroic past. It can also be read, I think, as nostalgia for a passing theory of the material world, the pre-seventeenth-century cosmos of elements and humors which rendered subject and world deeply interconnected and saturated with meaning.[32] Gillies has argued that this saturation of meaning, "a rich geographic tradition which is already moralized, already inherently 'poetic' in the sense of being alive with human and dramaturgical meaning," shapes Shakespeare's representation of marginal, outlandish, barbarous, and exotic non-European cultures as in need of control by the rational and self-controlled West.[33] Raman, on the other hand, links proto-colonialist representations of India and the East in Shakespeare's time with developments in geometry and cartography that led to "a changed understanding of space" and "a Western man, adequate to that space."[34] Marlowe's *Tamburlaine* uses cartography in just this way.

Certainly, the beliefs and prejudices inherited from classical antiquity and the technologies produced by the "new science" contributed to the ideologies that justified European colonial domination of India, Africa, and the Americas.[35] However, the question raised by Shakespeare's *Antony and Cleopatra* is less *which* of these two modes of thought is the most implicated in colonialism than *how* each belief system works as a mode of inhabiting the world. Although Egypt in the play is certainly marked by orientalizing stereotypes, it also represents a mode of thought that Shakespeare associated with sixteenth-century England and that is experientially resistant to a will to colonize others or, finally, to being colonized itself.

Although the imperial Roman paradigm wins out in the historical narrative traced by the play, the Egyptian mode proves itself to be difficult to fix, pin down, or grasp. Its intangibility, imagistic richness, and extravagance ally it with the poetic and the theatrical—that is, with literature as opposed to "science" in its modern sense. The play thus traces the nascent split between C. P. Snow's "two cultures," although it may also suggest that the Egyptian relation to the natural world involves a kind of knowledge different from, but not necessarily inferior to, the scopic economy of the new science.[36]

"Earth" in *Antony and Cleopatra* is both another name for the "world" and the name of one of the constitutive elements of Aristotelian science: earth, water, air, and fire. Antony's early declaration of love for Cleopatra clearly articulates the difference between imperial world and material earth; "Let Rome in Tiber melt, and the wide arch / Of the rang'd Empire fall! Here is my space, / Kingdoms are clay; our dungy earth alike / Feeds beast as man" (1.1.33–36). The Roman world is an orderly, impermeable, man-made "arch." The Egyptian "earth" is "dungy" "clay"—elemental, life giving, and allied with another element, water. In general, characters associated with Egypt perceive the world as composed of the four elements. Gillies notes the prevalence of water imagery in the play, but the other elements are present as well.[37] The soldiers who believe they hear Hercules abandoning Antony perceive mysterious "music i'th' air" or "under the earth" (4.3.13). While the Romans refer to battles waged on "sea" or "land" rather than water and earth, Antony reveals his mixed allegiance; he notes that the Roman "preparation is to-day by sea, / We please them not by land," while he and his Egyptian allies "would they'ld fight i'th' fire, or I'th' air" (4.11.1–3). Cleopatra heralds her coming death when she proclaims, "I am fire and air, my other elements / I give to baser life" 5.2.289–90). In addition, Aristotelian science posited a connection between the elements constructing the macrocosm and the humors that composed the human microcosm, such that Earth and its inhabitants were made of the same interchangeable stuff. In attempting to account for changes in nonliving elements, Aristotle drew an analogy from the operations of the human body (digestion, concoction, etc.) to understand the operations (melting, evaporation, ripening, decay) of nonhuman matter. In this universe, people and the Earth were inextricably intertwined.[38]

Bruno Latour has emphasized the "work of *retrofitting* that situates a more recent event as what 'lies beneath' an older one, " for instance, reading back from our knowledge of experimental evidence for the existence of atoms that was produced in the early twentieth century to recognize the speculative atomism of the seventeenth century as "real science." Latour's insight reminds us that in 1607, the "new" science might have seemed to be separating the tangible surfaces of the world from their invisible material underpinnings, even though experimental evidence later caught up with theory and retroactively provided underpinnings for it.[39] As we have seen, the end of belief in the Aristotelian elements meant the end of a system in which human sensory experience of the world was thought to give unmediated access to truth about it and in which humans and the world were interconnected in complex ways. Instead, the new science fostered a system in which visual observation, categorization, and naming of the surfaces of matter placed rational man above and apart from the world and (in theory) enabled him to control it. Elizabeth Spiller has argued that the transition from Aristotelian theory to the new science was mediated by a new sense, derived from the mechanical arts, that artificial means could yield knowledge of and control over nature.[40] In *Antony and Cleopatra*, Shakespeare explores what it feels like to move from a natural, elemental world to an artificially constructed one.

The play reflects this transition when it depicts Egyptians as permeable and interconnected with their elemental Earth and Romans as dominating a hard-surfaced, impervious world. The Egyptian worldview seems to reflect what Mary Floyd-Wilson has termed "geo-humoralism," the pervasive early modern belief that climate and other environmental factors shaped the bodily complexion of humors and, therefore, shaped racial character.[41] The point that I want to emphasize here is not that Egyptians seem stereotypically "warm-blooded" and self-indulgent, while Romans are cold, austere, and self-controlled, but rather that Egyptians in the play reflect an earlier view that environment shapes subjects while the Romans look forward to a Cartesian mind-body split, in which self-contained individuals are separate from and gain mastery over their environment. As Floyd-Wilson argues, the development in the seventeenth century of a new ethnography based on a "disavowal of both environmental and somatic influences on the mind" allows for "the formation of the autonomous—and white—subject." Furthermore, "the construction of bounded selves goes hand in hand with the construction of racial boundaries."[42]

Thus, the Egyptians in the play imagine themselves as being fed by their "dungy earth," as Cleopatra describes it, "the dung, / the beggar's nurse and Caesar's" (5.2.7–8). Cleopatra imagines herself as the embodiment of Egypt because she has been shaped by its environment; she is the "serpent of old Nile," fed with the serpent's "delicious poison," who has become "with Phoebus' amorous pinches black" (1.5.25). Cleopatra imagines the triumph that Caesar plans for her in Rome in terms that emphasize the humoral and meteorological interrelation of body and environment. The Romans citizens have "thick breaths, rank of gross diet"; in their breaths Cleopatra will be "enclouded / And forced to drink their vapour" (5.2.207–8).

Romans, on the other hand, imagine themselves as impervious to environmental influence. They tend to conceive of the world as hard and solid, and they think of themselves as statues or buildings. Philo calls Antony "the triple pillar of the world." A messenger describes Octavia to Cleopatra as "a statue" (3.3.21). Caesar speaks of the "three-nooked world" (4.6.5), as if the human demarcation of Europe, Africa, and Asia determined its shape, and Menas calls the triumvirs "world-sharers" (2.7.70), as if they together possess the world. As a solid object (rather than fungible "dung" or "clay"), the Roman world is imagined as being broken in half by political division. Octavia believes that should there be war between Antony and Caesar, it would be "as if the world should cleave" (3.4.31), and Caesar says (depending on which text you use) that either the "round world" (5.1.15, *Riverside*) or "rived world" (5.1.15, *Oxford*) should be altered by Antony's death. In the Roman world, human relationships are imagined as "cement" holding parts of the world together, as when Pompey notes that Antony and Caesar will "square" between themselves when "fear of us/May cement their divisions" (2.1.45, 480). Similarly, Caesar imagines Antony's marriage to Octavia as a "hoop should hold us staunch from edge to edge/O'th world" (2.2.120–21).

Caesar's long, strange speech in Act 1, scene 4 (ll. 55–71), describing the trials that the formerly heroic Antony has survived, offers a weird fantasy that idealizes a complete imperviousness to the environment. Antony "didst drink the stale of horses," "brows'd" tree bark, and "didst eat strange flesh/Which some did die to look on" but suffered no bodily effects: "thy cheek/So much as lank'd not." Janet Adelman reads this speech as manifesting a "contest between Caesar and Cleopatra, Rome and Egypt" that "is in part a contest between male scarcity and female bounty as the defining site of Antony's masculinity."[43] Adelman is certainly correct that Rome and Egypt are strongly marked by gender difference. I would shift her emphasis a bit to note that women's bodies, as Gail Paster has shown, were seen in this period as more open to environmental influence—more porous, leaky, and impressionable—than male bodies and therefore that the gendering of Rome and Egypt in the play can be seen as complementary to the different relation of each to the material world.[44]

The Egyptian Earth is controlled not by its human inhabitants but by cycles of natural change and transformation that extend from birth through death. The flooding of the Nile and the fertility that it engenders is the central Egyptian trope: "The higher Nilus swells,/The more it promises; as it ebbs, the seedsman/Upon the slime and ooze scatters his grain,/And shortly comes to harvest" (2.7.20–23). This fertility is imagined in elemental terms, as the water of the Nile combines with earth to form "ooze" in which the sun engenders life: "By the fire/That quickens Nilus' slime" (1.3.68). "Ooze," "fire," and "slime" conjure up the feel of these materials, suggesting a sensory immersion in the elements rather than visual mastery of them. The interaction of these elements can even spontaneously generate life, as was widely believed until well into the seventeenth century: "Your serpent of Egypt is bred now of your mud by the operation of your sun. So is your crocodile" (2.7.26–27). Antony reiterates this belief that a "courser's hair" could be transformed into a snake that "hath yet but life/And not a serpent's poison" (1.2.176). Cleopatra imagines that a kind of reverse generation is also possible when she exclaims, "Melt Egypt into Nile! and kindly creatures/Turn all into serpents" (2.5.78–79). And, although she can at one point imagine her death as a transformation into immaterial "fire, and air," she can also imagine it as a return to the ooze of the Nile: "Rather a ditch in Egypt / Be gentle grave unto me! Rather on Nilus' mud / Lay me stark-nak'd, and let the water-flies / Blow me into abhorring!" (5.2.57–60) In Egypt, characters feel themselves to be part of the processes of nature upon which they depend and which they can't control.[45]

Romans, on the other hand, view the world as changed only as a result of human agency. They don't seem to perceive the natural cycles that so shape Egypt, nor do they imagine themselves as part of them. Romans rely on visual observation of a world that they almost obsessively divide into geographic entities, which they name and control by naming: "Labienus— / This is stiff news—Hath with his Parthian force / Extended Asia; from Euphrates / His conquering banner shook, from Syria / To Lydia and Ionia" (1.2.88–92). In *Space, Geography, and Politics in the Early Roman Empire,* Claude Nicolet describes how, "in order to set boundaries to their empire and to claim to have reached those that were marked out, the Romans needed a certain perception of geographical space, of its dimensions, and of the area they occupied."[46] He charts the means through which Romans established this perception of space, some of which have remarkable resonance with Shakespeare's play. In the *Res Gestae* of Augustus, for instance, lists of geographical names ("over fifty-five geographical names divided into four large categories") play a role in helping Romans imagine the world as a space that they could control.[47] Similarly, he notes the "ever-more-frequent appearance of the globe on Roman coins from about 76 B.C. or 75 B.C.," a representation that made the "world," or "orbis terrarum," visible as an artifact.[48] Shankar Raman has linked this moment of imperial Roman domination of space with an "analogous" early modern moment of "the material domination and symbolic appropriation of space," when "staging a geometrized and neutralized space . . . helped conceal the colonial practices out of which they emerged and to which they contributed."[49]

The separate, solid, man-made, nameable world in Shakespeare's play thus reflects spatial strategies necessary for imperial domination. When Caesar wishes to portray Antony and Cleopatra as a threat to Roman imperium, he makes them sound like Tamburlaine when he describes them engaged in a scene of imperial naming:

Unto her [Cleopatra]
He gave the stablishment of Egypt, made her
Of lower Syria, Cyprus, Lydia,

Absolute queen . . .
Great Media, Parthia, and Armenia
He gave to Alexander; to Ptolomy he assign'd
Syria, Cilicia, and Phoenicia. (3.6.8–11, 14–16)

Caesar imagines the scene as ornamented with orientalizing trappings: "I' th' market-place, on a tribunal silver'd;" "in chairs of gold," "She/In th' abiliments of the goddess Isis/That day appear'd" (3.6.3, 4, 17–18). Although he attributes these gestures of empire to Antony and Cleopatra, we never see them engage in anything like the scene he imagines, and it seems possible that he fabricates or exaggerates it in order to justify waging war against Antony. It is significant that Caesar translates the earthy imagery of Egypt into hard, though exotic, surfaces—a silver tribunal and a golden chair, as if he cannot imagine Egypt on its own terms.

Egyptian earth, on the other hand, is less suited for conquest—that is, either to conquer or to be conquered. Shaped by their environment, mired in the ooze of the Nile, seemingly inseparable from the earth that gives birth to them and receives their dead bodies, Shakespeare's Egyptian subjects lack the objectified concepts of space and geography that lead to *imperium*. In the scenes where battles between Antony's Egyptian forces and Caesar's attacking troops are discussed and described, Cleopatra's influence is repeatedly associated with a refusal to occupy and defend hard ground, the space on dry land upon which the Roman world is based.[50] Instead, she insists on fighting at sea, and her sudden retreat is associated with the yielding elements of water and air; "the breeze upon her, like a cow in June—/Hoists sails and flies; " "Our fortune on the sea is out of breath;" (3.10.14–15, 24). Antony's reaction to this first loss at sea takes the form of a sense that the land has rejected him: "the land bids me tread no more upon't . . . I am so lated in the world" (3.11.1, 3). After Cleopatra's second retreat at sea, Antony imagines the world and, ultimately, himself as made of the yielding and indistinct elements of water and air that make up clouds and mist.

Sometime we see a cloud that's dragonish,
A vapor sometime like a bear or lion,
A tower'd citadel, a pendant rock,
A forked mountain, or blue promontory
With trees upon't that nod unto the world
And mock our eyes with air. (3.14.2–7).

These seemingly solid and visible shapes are as insubstantial as his political power and control of Egypt has become: "even with a thought/The rack dislimns, and makes it indistinct / As water is in water . . . now thy captain is / Even such a body" (3.14.11–13).[51]

While the Egyptian forces are unable to achieve mastery, visual or otherwise, over the solid world, we must nonetheless wonder whether Caesar really is able to conquer Egypt in any meaningful sense. Gil Harris cites Enobarbus's famous description of Cleopatra to argue that although "Cleopatra's power appears to be predicated on the visibility of her eroticized body to her subjects, who abandon all activity to gaze on her," Enobarbus never actually describes *her*: "For her own person, / It beggar'd all description."[52] Harris cites the "synaesthesic" nature of the description of her surroundings, but it may be more accurate to say that it simply focuses on senses other than sight: the sails "so perfumed that / The winds were lovesick with them;" the oars keep time to "the tune of flutes;" boys wield fans "whose wind did seem / To glow the delicate cheeks which they did cool;" her gentlewomen tend the tackle with "flower soft hands;" the barge gives off a "strange invisible perfume" (2.2. 193, 195, 203–4, 210, 212). Cleopatra's Egyptian power manifests itself as a spectacle that cannot be fully seen and therefore cannot be captured by sight, the Roman vehicle of mastery and empire.

Enobarbus's speech thus eschews the Egyptians' earthy landscape of mud, slime, and ooze, replacing it with an ethereal fantasy that is equally ungraspable. This is a vision of fire and air rather than earth and water, a "vision" that is not quite seeable, since Cleopatra is insistently likened not to concrete objects but to her differences from them. She "o'erpictures" a portrait of Venus, and her accouterments are difficult to attach definitively to a solid surface: What exactly is made of "cloth of gold of tissue"? To what is the "silken tackle" attached? How does the tackle "swell"? The culmination of this vision is, appropriately, the threat of "vacancy" or vacuum, not any concrete presence.[53] If Egypt can't be clearly seen or firmly touched, it seems difficult to know or conquer it with any certainty.

Cleopatra temporarily adopts Roman language when she falsely assures Caesar that "all the world; 'tis yours, and we, / Your scutcheons and signs of conquest, shall / Hang in what place you please" (5.2.134–36), but her suicide prevents him from leading her in triumph through Rome as a sign of his conquest. Although his final speech attempts to monumentalize Antony and Cleopatra in Roman terms—"no grave upon the earth shall clip in it / A pair so famous," and "our army shall / In solemn show attend this funeral"—many readers have felt that his gesture is inadequate as a final word and agree rather with Cleopatra's conclusion that "'Tis paltry to be Caesar."[54] Thus, although Caesar does materially conquer Egypt, it is not clear that he has gained any purchase on its way of life. As Ania Loomba has argued, "Cleopatra's final performance, which certainly exposes her own vulnerability, not only cheats Caesar but denies any final and authoritative textual closure" (130). Indeed, although Egypt was annexed by the Roman Empire after the Battle of Actium in 31 BC, Alexandria remained a center of learning to rival Rome and never really adopted Roman customs. Under the emperor Constantine, the center of the Roman Empire was moved east, to Byzantium, and Alexandria remained an important center of learning after the Byzantine Empire replaced the Roman.

Although I've been arguing that Egypt, as depicted in the play, represents something other than (or more than) an orientalized stereotype, it is important to note that the Egyptian worldview is also flawed. As Gil Harris has argued, Cleopatra sometimes seems curiously disembodied despite the insistent corporeality of her language, since "reminders of her physicality are supplemented by a counter-narrative in which her very vividness is shown to be the effect of a Roman desire for her presence, prompted by the gaps and absences that repeatedly afflict the play's attempts to represent her."[55] It's as if the play can't quite believe in the elemental Egyptian earth and transcendent fire and air but at least entertains the idea that they are a powerful fantasy. Thus, Enobarbus's description of Cleopatra seems like a fantastic dream, and the clearest articulation of the central Egyptian trope of spontaneous generation comes from the drunken Lepidus, who desires to believe in an exotic Egypt that he has never seen.

If spontaneous generation is the central trope of Egyptian earthiness, then Egypt is linked to the complex and problematic history of this concept, which raises some of the central issues of the play. Aristotle offers different accounts of spontaneous generation, reflecting the problems that it entailed for his theories of causation and matter. From the beginning, the process of spontaneous generation raised questions about the nature of the material world. Was matter wholly natural, or was it infused with some sort of divine spirit? Could life itself be explained in wholly material terms, or did it require a divine spark? In *The History of Animals,* Aristotle describes spontaneous generation as a wholly material process but gives no explanation of how it works: "So with animals, some spring from parent animals according to their kind, whilst others grow spontaneously and not from kindred stock; and of these instances of spontaneous generation some come from putrefying earth or vegetable matter, as is the case with a number of insects, while others are spontaneously generated in the inside of animals out of the secretions of their several organs."[56] In *The Generation of Animals,* book 3, chapter 11, however, he considers how life could be spontaneously generated from nonliving elements and concludes that a life spark or "pneuma" must permeate matter and can cause life under the right circumstances.

> Animals and plants come into being in earth and in liquid because there is water in earth, and air in water, and in all air is vital heat so that in a sense all things are full of soul. Therefore living things form quickly whenever this air and vital heat are enclosed in anything. When they are so enclosed, the corporeal liquids being heated, there arises as it were a frothy bubble. Whether what is forming is to be more or less honourable in kind depends on the embracing of the psychical principle; this again depends on the medium in which the generation takes place and the material which is included.[57]

Doubts about the truth of spontaneous generation surfaced early in antiquity and coexisted with acceptance of it as a fact well into the seventeenth century. Theophrastus questioned it, as did Lucretius. In the early seventeenth century, the Paracelsan alchemist and physician J. B. van Helmont "believed that frogs, slugs, and leeches were spontaneously generated" and provided directions for the generation of mice: "if a dirty shirt is stuffed into the mouth of a vessel containing wheat, within a few days, say 21, the ferment produced by the shirt, modified by the smell of the grain, transforms the wheat itself, encased in its husk, into mice."[58] William Harvey also accepted spontaneous generation as a fact. It wasn't until the 1660s that Francesco Redi demonstrated that maggots did not appear spontaneously in rotten meat and not until the nineteenth century that Louis Pasteur disproved the spontaneous generation of microbes.[59] The history of the theory of spontaneous generation thus reflects the epistemological shift enabled by increasing skepticism about intuitive science that I have been emphasizing here. Aristotelian science took visual evidence that lower forms of life seemed to appear without visible cause in various media to mean that they must be spontaneously generated, since human sense perception should be able to detect the material causes of these life forms if they indeed existed. Latour suggests how difficult it was for Pasteur to eradicate "the well-known universal phenomenon of spontaneous generation," which could only be done through "a gradual and punctilious *extension* of laboratory practice to each site and each claim" of its defenders.[60] With each stage and through great effort, experimentation and technology move scientific theory farther from the purview of ordinary human perception; with each stage, too, matter loses its animate spark, and living things are more strictly separated from nonliving things.

Shakespeare's play insists on belief in spontaneous generation as a central feature of Egyptians' relation to the Earth. Only the credulous and drunken Lepidus among the Romans seems to entertain the theory. An Earth with the capacity to generate life is seen as nearing the end of its tenure, to be paved over by Romans who free themselves from enmeshment with the elements by constructing and colonizing an inert, nonliving world. Although not yet envisioning the mechanistic universe that would become dominant by the end of the seventeenth century, the Romans do imagine an artificial world. The play does not attempt to judge which worldview is scientifically correct. It simply marks the passing of one into the other and registers the perceptual experience of each.

Although Rome (and science) triumph in the temporal space of the play, the copiously productive Egyptian earth provides a more fertile source for Shakespeare's poetic imagination, with its proliferation of metaphors and analogies, than the spare and plain style of the Roman world. In constructing Rome and Egypt as a perceptual dichotomy, the play may mark an originary site of the disciplinary division into the two cultures of literature and science which has so deeply structured modernity. The play pauses for a moment to register the loss of an intuitive connection with the Earth. The nostalgia that seems to attend the final scenes of the play may, in fact, reflect the passing of a worldview that lent itself more readily to the Shakespearean imagination in all its abundance and ambivalence. The play seems to acknowledge the greater efficiency of the Roman mode, as well as its greater potential for domination of the world and its inhabitants, even as it acknowledges what literature will lose as a result.

What about Bacon?

Almost every early modern scholar to whom I have described the scope and argument of this book has asked, "What about Bacon?" Their assumption has been that a book about the epistemological background of the scientific revolution would necessarily have to discuss Francis Bacon, who is widely perceived as having facilitated the development of an empirical and inductive approach to scientific discovery, inspiring the foundation of the Royal Society and making possible a systematic approach to the new science. Bacon was a controversial figure during his life, and his reputation among scholars has been equally contentious. I think it is safe to say that literary scholars generally hold him in higher esteem than do historians of science and philosophy, many of whom have questioned whether his program for the advancement of science represented real innovation and whether it provided a workable scientific method.[1]

Placing Bacon in the context of the sixteenth-century writers discussed thus far in this book provides new insight into his work and the nature of his achievement. On the one hand, Bacon was explicitly responding to the impasse at which English natural philosophers found themselves at the end of the sixteenth century. If intuitive knowledge of the universe based on ordinary sensory experience had proven to be unreliable in many respects, on what basis could the study of nature move forward? As we have seen, advocates of various "secrets" traditions described technologies that they hoped would provide access to the real truth hidden under deceptive appearances. Bacon's new inductive method can be seen as another one of these technologies, designed to bridge the gap between flawed observation and underlying truth. Bacon's work differs from that of writers like Thomas Digges, Thomas Harriot, and William Gilbert in a number of ways, however. His system is not arcane and secret but finds a way to make the process of attaining hidden knowledge into an open and communal undertaking, as well as to make that process itself, rather than the truth about nature that it discovers, seem intuitive. Bacon's project is more ambitious, claiming to provide a systematization of all knowledge, not just natural philosophy. Bacon's system is not based in the methodologies of mathematics or astronomy or optics; instead, as Paolo Rossi and Lisa Jardine have shown, he expands the techniques of humanist dialectic to provide a program for the invention (or discovery) and judgment (or analysis) of knowledge of all kinds.[2] And, as Barbara Shapiro has argued, he imported the concept of "fact" from law and history to the realm of natural philosophy.[3]

Ultimately, though, Bacon's main innovation lies in how he felt about the loss of the Aristotelian consensus and how he persuaded others to feel about it. Rather than regretting the loss of this worldview, he vehemently attacked it. He saw its decline as an opportunity for advancement rather than a sign that the world might be about to end. He replaced anxiety and doubt with optimism and measured confidence. Like Gabriel Harvey (whom he resembles in a number of ways), Bacon was able to view the epistemological crisis of the late sixteenth century without anxiety because he did not fully understand the implications of the discoveries of Copernicus, Digges, and others. Unlike most sixteenth-century natural philosophers, who downplayed the originality of their ideas and claimed ancient precedents for them, Bacon exaggerated the innovative nature of his program and ignored or dismissed his ancient and modern precursors.

We can blame Bacon, in fact, for many of the misconceptions about sixteenth-century natural philosophy that I described in the introduction. Bacon was vociferous in his rejection of Aristotle, but he ignored the naturalistic texts and focused his critique on the Aristotle of university disputation, his rules of syllogistic reasoning in particular.[4] In the *Novum Organum*, Bacon argues that Aristotle "with his dialectic corrupted natural philosophy when he fashioned the world from categories," with the result that "*Aristotle's* Physics rings out with little more than the terms of dialectic."[5] He seems aware of the naturalistic tradition but dismisses it: "And no one should be impressed by the fact that in his books on Animals, in his Problems, and in other tracts of his he often deals in experiments" (*Novum Organum*, 101). According to Bacon, Aristotle approached the observation of nature with his ideas already formed and then "bends experience to his opinion and drags it about in chains" (101). Bacon doesn't recognize the important distinction between "experience" and "experiment" and doesn't acknowledge contemporary work like that of Digges or Gilbert which understood the distinction and engaged with its epistemological implications. His text relies instead on the powerful image of Aristotle subduing a personified Nature and dragging her along in chains, emphasizing rhetorical effect over methodological precision.

Bacon, interestingly, does not mention Aristotle's *Meteorology*, although his *History of the Winds* is heavily based on its assumptions. Bacon asks there, for example, "What do meteors of different kinds contribute to the winds? What do earthquakes, rains, and the coming together of the winds, contribute? For these things are linked together, and one draws along the other."[6] In *The Advancement of Learning*, Bacon accuses Aristotle (as well as Galen and Plato) of studying formal and final causes rather than the more empirically accessible material and efficient causes, ignoring the fact that the meteorological tradition focused on material and efficient causes and that contemporary meteorology (in works by Harvey and William Fulke, for instance) included sophisticated discussion of the relationship between final and material causes. Instead, Bacon accuses Aristotle and Galen of considering only final causes, of arguing that "*clouds are for watering of the earth*," and of inhibiting scientific discovery because "the physical causes hath been neglected, and passed in silence."[7] *Meteorology*, of course, includes extended discussion of the physical causes of clouds and other phenomena.

We can also trace to Bacon the indiscriminate dismissal of most existing "secret" traditions as superstitious and purposefully obscure. He lumps together "astrology, natural magic, and alchemy," conceding that they have "noble ends" and praising the "search and stir" of alchemy which has "brought to light a great number of fruitful inventions and experiments." In practice, though, these pursuits are "full of error and vanity," accepted only because people are too credulous.[8] Later in the *Advancement* he negatively characterizes natural magic: "the Natural Magic whereof now there is mention in books, contain[s] certain credulous and superficial conceits, and observations of sympathies and antipathies, and hidden properties, and some frivolous experiments, strange rather by disguisement than in themselves."[9] Rather than attending to the useful work in fields like optics that contributed to natural magic in the sixteenth century, Bacon focuses on the most extreme claims of writers like Paracelsus and argues that all work in natural magic makes the same spurious claims.

Bacon accuses Aristotle of being too critical of rival philosophical systems: "he never nameth an ancient author or opinion, but to confute and reprove."[10] This is, of course, true of Aristotle but also true of Bacon himself. Unlike the members of the Dee and Northumberland circles, who ready widely in ancient, medieval, and contemporary natural philosophy in search of useful ideas, Bacon tends to emphasize what is wrong with every other approach, even those of writers like

Ramus and Paracelsus whose ideas deeply influenced his own. He dismisses William Gilbert's work, although Gilbert represented one of the best contemporary examples of someone gathering experimental data in a Baconian fashion, placing Gilbert among those who "infect their meditations, opinions, and doctrines with some conceits which they have most admired, or some science which they have most applied." Gilbert, according to Bacon, "made a philosophy out of the observations of a lodestone."[11] Bacon's discussion of Copernicus suggests that he was as confused as Harvey or Richard Bostocke about how to view his cosmology. Bacon claims that the Ptolemaic and Copernican systems are equally valid mathematically, something that Robert Recorde and Digges knew to be untrue, and Bacon's own cosmology assumes that the earth is stationary.[12]

Graham Rees provides a clear account of Bacon's theories about the universe, what Rees calls his "speculative philosophy" viewed separately from his discussion of scientific methodology. Rees rather harshly judges Bacon in this regard, arguing that he "raided disparate natural philosophical traditions for attractive titbits which he refashioned and spatchcocked together to form a curious hybrid which embodied, even by the standards of the early seventeenth century, some very peculiar alliances of ideas indeed."[13] Bacon's cosmology is probably not any more "peculiar" than the one developed by John Dee in his later years, but it does combine elements from traditions that were not really compatible; Rees mentions Aristotle, Copernicus, Galileo, Paracelsus, William Gilbert, Telesio, and Alpetragius.[14] Differing sharply from mainstream views, Bacon seems to have divided the universe into three zones: the core of the earth, the heavens, and a border zone in between, all composed of different materials. He believed that the heavenly bodies rotated around the Earth from east to west, at different speeds depending on their location (and therefore their chemical composition). Rees argues that Bacon espouses these "principles which few other European philosophers took seriously" because he realized that the Copernican system was "gaining ground" on the Aristotelian system among many European thinkers but could not bring himself to accept a Copernican theory that "offended common sense."[15]

Despite its eclecticism, Bacon's theory of the universe is in many ways more conservative than the theories of Digges and Gilbert. Bacon's cosmology avoided the most threatening deviations from Aristotelian natural philosophy and comprised, in Rees's words, "a finite, geocentric, plenum," so the notions of infinity and void space which troubled Shakespeare do not apply. Bacon also maintained a division between sublunar and supralunar physics, although he adapted a three-zone structure from Telesio and borrowed other ideas about the makeup of the three zones from Alpetragius and Paracelsus.[16] He does not seem to have been aware of the significance of the location of the supernova of 1572 and posited a supralunar matter more "pure and enduring" than that of the sublunar realm.[17] Rees also emphasizes that Bacon's natural philosophy was "bounded" by theology and that theological truths "placed limits on choice" for him in matters of natural philosophy.[18]

Bacon seems to have been uninterested in the problems that other sixteenth-century English natural philosophers found pressing: precession, retrogradation, the nature of the fixed stars, the behavior of light. This is partly the case because he did not believe that answers to these puzzles could be found without a complete and systematic revision of the method of natural philosophical inquiry. Bacon's Aristotle, the Aristotle of the university arts curriculum, claimed to offer a system for the achievement of logical certainty, while in practice, at least in Bacon's opinion, it offered only complicated rules for the construction of trivial arguments. He accused "the schoolmen" and "Aristotle, their dictator," of practicing "contentious learning," in which "it is the property of good and sound knowledge to putrify and dissolve into a number of subtle, idle, unwholesome, and as I may term them, vermiculate questions."[19]

Bacon looked instead to the revised dialectics of Agricola, Melanchthon, and Ramus; these were systems for "inventing" (finding) arguments and then for "judging" (choosing and arranging) them and were used in teaching students Latin composition and basic methods of disputation. Bacon further revised and extended these techniques so that they could be applied to the discovery and analysis of observations from nature rather than just arguments from books.[20] His inductive method thus resembles the "commonplace book" method taught in early modern schools because it advocates first gathering information in the form of natural history, then judging or analyzing what has been gathered in order to build toward general axioms.[21]

Optics, for Bacon, was not a source of experimental and quantifiable knowledge about the relationship between appearance and reality, and he did not share contemporary interest in the potential importance of optics in astronomy or meteorology. It may be that Bacon believed that the fallibility of human sense perception could be remedied by proper inductive method, so better understanding of vision itself was not important to him. Optics did, however, provide him with a fruitful source of metaphor. I would argue that his reliance on optics as a trope rather than an experimental science is a product of the basis of his method in dialectics. In the *Advancement*, for instance, Bacon identifies three "beams" of human knowledge: "radius directus," or nature; "radius refractus," or God; and "radius reflectus," or man's knowledge of himself.[22] Unlike Harriot's exchange with Kepler on the reflection and refraction of light through media of varying densities, Bacon uses the observed behavior of light as a rhetorical lynchpin of "method," a metaphorical basis for a division of knowledge into three constituent parts.[23]

Bacon most closely resembles other late-sixteenth and early seventeenth century natural philosophers in his sense that new technologies were needed to bridge the gap between ordinary sensory experience of nature and deeper truths about it. He uses another optical metaphor when he explains that "the mind of man is far from the nature of a clear and equal glass, wherein the beams of things should reflect according to their true incidence; nay, it is rather like an enchanted glass, full of superstition and imposture, if it be not delivered and reduced."[24] Unlike Digges or Dee or Gilbert, though, Bacon believed that the mind's reflection of nature was inaccurate not because nature itself was deceptive but because of human "superstition and imposture," which could be corrected by using the proper method of inquiry. Bacon was more optimistic than many of his contemporaries about the adequacy of the senses to access truth, asserting that the senses are "sufficient to certify and report truth though not always immediately, yet by comparison, by help of instrument, and by producing and urging some things that are too subtile for the sense to some effect comprehensible by the sense, and other like assistance."[25] He avoids the Aristotelian term *occult* as a way of naming what is "too subtile for the sense," preferring to use "latent," perhaps because *occult* was already taking on its modern connotation of "supernatural," and Bacon wanted to differentiate his program from what he considered "superstitious" secrets traditions.[26]

Bacon's "Idols" include the idea that intuitive knowledge is, in and of itself, an impediment to knowledge. He makes intuitive science sound too easy to be true, especially when contrasted with the rigors of his inductive method. The "Idols of the Tribe," which characterize the innate shortcomings of the human mind, include the fact that "the human intellect is swayed most by those things that can strike and enter the mind suddenly and in one go" and is less inclined to "rapid recourse to the remote and heterogeneous instances by which axioms are tested as if by fire."[27] The "greatest hindrance" to understanding is that "things which strike the senses outweigh things which, even if they are more important, do not strike them immediately," so that "things invisible attract little or no attention."[28] Bacon suggests that we should be suspicious of knowledge that offers itself up immediately and without effort, and he insinuates that reliance on intuition is lazy, in contrast with the energetic "rapid recourse" of his own approach to knowledge.

The language with which Bacon describes his "instrument" for repairing the failure of intuition both resembles and differs from that of the writers I have been discussing in this book. While a Paracelsan like Bostocke argued that it was only possible to penetrate the obscuring cover of visible qualities through "the fire," using alchemical techniques of distilling and refining, Bacon explicitly substitutes induction: "thus we must make a separation and dissolution of bodies not by fire indeed but by reason and true *Induction* with experiments to reinforce them; and by comparing them to other bodies, and reducing them to simple natures and their forms, which in a compound body come together and become intertwined."[29] Bacon substitutes for physical separation of substances in a laboratory the intellectual division of concepts into "simple natures."

Bacon's concept of "form" had an important role in his program for finding deeper truths behind the appearances of nature. Antonio Pérez-Ramos summarizes "the near universal verdict" that Baconian Forms were "a totally useless tool in the construction of modern science."[30] However, he goes on to trace the importance of "Form" for Bacon as a way to bridge the gap between perceptible qualities and the underlying structures that produce them. As Ramos describes the process, "if there is a given nature which we perceive, then there should be a Form of it which sense cannot attain but whose construal is the work of the intellect governed by true method, that is, by induction."[31] Bacon could not quite accept the theory that atoms made up this invisible formative substrate of matter, so the concept of Form allowed him to offer a method for revealing the substrate without committing himself in advance to its nature.[32]

We can recall that late sixteenth-century natural philosophers tended to describe whatever method they advocated for revealing hidden knowledge as offering a means—wings, ladder, stairs—of providing access to a previously inaccessible realm. We saw Digges claiming to provide "wings or ladders" up to the stars, while Gilbert offered a way to travel to the depths of the Earth. In a revealing passage in the *Advancement*, Bacon picks up on this language but changes it. He cites Democritus and "the alchemists" for the idea that knowledge is "hid in certain deep mines and caves" and that fire can provide access to what is hidden. Bacon typically transforms these tropes into the conceptual basis for a dialectical partition into categories, arguing that "it were good to divide natural philosophy into the mine and the furnace." Miners, in Bacon's view, would "dig" in order to make "inquisition of causes." Some effort is needed to access hidden causes, but it does not involve an arduous journey into a previously hidden realm, and this mining operation isn't, in itself, adequate to reveal the truth. Further analysis is needed. The fire thus becomes an agent not of the alchemist's laboratory but of the blacksmith's forge. Natural philosophers who aren't miners will "refine and hammer" what the miners dig up, leading to a practical "production of effects."[33]

Unlike the ladder into the stars offered by Digges, Bacon argues that the ladder of knowledge must go in two directions: "all true and fruitful natural philosophy hath a double scale or ladder, ascendant and descendant; ascending from experiments to the invention of causes, and descending from causes to the invention of new experiments."[34] Bacon makes the promise of a one-way trip into the heavens sound like a frivolous vacation. What is needed, instead, is a systematic and laborious undertaking, repeatedly climbing up and down the ladder of induction. In general, Bacon likes to emphasize that his method is more serious and difficult than those of his rival philosophers. Instead of the superficial cosmologies of Ptolemy and Copernicus, "the real truth requireth another manner of severity and attention."[35] His method is preferable to that found in contemporary treatises of natural magic because "whoever shall entertain high and vaporous imaginations, instead of a laborious and sober inquiry of truth, shall beget hopes and beliefs of strange, impossible shapes."[36]

Bacon uses a metaphor familiar from humanist education to convey the laborious but transformative power of true induction. From classical antiquity, the image of a bee gathering nectar from flowers and distilling honey from that nectar was used to describe the process of reading, assimilation, and imitation of previous authors. Erasmus sums up the commonplace book method of invention in this way: "our student will flit like a busy bee through the entire garden of literature, will light on every blossom, collect a little nectar from each, and carry it to his hive."[37] Bacon reveals the sources of his inductive method in techniques of dialectical invention, calling on the image of the bee when he differentiates his method from the "utterly vicious and incompetent" induction taught by ancient authors. True induction seeks the "dew of knowledge" by "distilling and contriving it out of particulars natural and artificial, as the flowers of the field and garden."[38] He elaborates the bee metaphor in the *Novum Organum*, again to differentiate his true method from that of "empirics," who, like ants, "only store up and use," and that of "rationalists," who are spiders making "webs from their own entrails." The Baconian bee "collects its material from the flowers of the field and garden, but its special gift is to convert and digest it."[39]

Despite Bacon's concern that intuitive ideas about nature were deceptively simple, images like the ladder and the bee make the process of Baconian induction seem intuitive. While proponents of "secrets" traditions emphasized the special and arcane nature of the processes needed to provide access to truth, Bacon suggests that induction is labor intensive but natural. The raw data is as obvious as the flowers of the garden or field, and the human mind can distill truth from it if the proper method is industriously applied. Reid Barbour has argued that Bacon kept coming back to atomic theories of matter, despite his suspicion that they were too simple, because he found them intuitively satisfying: "for Bacon, atomism facilitates and comforts the beleaguered pioneer in the wilderness of subtle particulars."[40] Jardine has argued that Bacon "apparently believed that essential truth is guaranteed by the initial procedure of intuitive selection of simple natures and adventitious conditions of essences from the preliminary natural history."[41] When Bacon identifies the "fundamental physical properties," such as "gravity and levity," "density," "tenuity," "heat," "cold," and so on, which comprise an "alphabet" that can be analyzed to provide access to forms of matter, he does not describe a process for deriving those fundamental properties. They are, in Pérez-Ramos's words, "posited a priori" on the basis of the philosopher's intuition.[42]

I am among those who believe that Bacon did not make much substantive contribution to real knowledge about the world. I admire the anxious searching of Recorde, Digges, Dee, and Harriot more than Bacon's boosterism for his inductive method, and I also blame Bacon's polemics for some scholars' tenacious misrepresentation of sixteenth-century natural philosophy. But there is no denying that Bacon had more impact on the development of the new science than any of the writers whose work I discussed in earlier chapters. In order to move forward, natural philosophy at the end of the sixteenth century needed to find a way to shift its focus, to turn away from the expectation that ordinary observation of nature yielded the truth, and to find a compelling way to articulate a program for searching out the invisible and the counterintuitive. Bacon's solution to this problem was to turn away from nature itself and toward the method by which truth about it might be discovered. He developed a powerful myth, that rigorous scientific method could guarantee accurate knowledge, and he made that method seem intuitively satisfying, although difficult. By focusing on method rather than on nature itself, he was able to foster optimism that empirical data could provide access to truth that the senses couldn't accurately perceive. Bacon's outsider status (he was a lawyer, not a member of any of the London mathematical or scientific circles) and relative ignorance, combined with his ambition and synthetic intelligence, let him accomplish this remarkable feat. By applying dialectical methods to natural philosophy, Bacon did apply humanist learning to reshape science. His skillful use of rhetoric helped him convince others to share his enthusiasm and optimism, if not the method itself.

Notes

CHAPTER 1: Introduction

1. See, esp., Steven Shapin and Simon Schaffer, *Leviathan and the Air-Pump: Hobbes, Boyle, and the Experimental Life* (Princeton: Princeton University Press, 1985); Peter Dear, *Discipline and Experience: The Mathematical Way in the Scientific Revolution* (Chicago: University of Chicago Press, 1995); Barbara J. Shapiro, *A Culture of Fact: England, 1550–1720* (Ithaca: Cornell University Press, 2000); Richard Westfall, "Science and Patronage: Galileo and the Telescope," *Isis* 76 (1985): 11–30; Lorraine Daston and Katharine Park, *Wonders and the Order of Nature: 1150 to 1750* (Cambridge: MIT Press, 1998).

2. See Pamela Smith and Paula Findlen, *Merchants and Marvels: Commerce, Science, and Arts in Early Modern Europe* (New York: Routledge, 2001). For the role of literature as a practical art, see Henry S. Turner, *The English Renaissance Stage: Geometry, Poetics, and the Practical Spatial Arts, 1580–1630* (Oxford: Oxford University Press, 2006), and Elizabeth Spiller, *Science, Reading, and Renaissance Literature: The Art of Making Knowledge, 1580–1670* (Cambridge: Cambridge University Press, 2004).

3. Carlo Mazzio, in her introduction to a special issue of the *South Central Review* on *Shakespeare and Science, c. 1600,* cautions against "a teleological drive toward the 'new science,' empiricism, or even post-Enlightenment science" (3) but does argue that we should attend to "the distinctly narrative, literary, and hypothetical dimensions of experimental science and, conversely, the scientific dimensions of the literary or fictional experiment" (4). She cites Howard Marchitello's argument that "literary culture is no longer believed to exist in a merely reflective relation to the disciplines of science; instead, science and literature are set in a creative dialectic with each other that denies priority and scientism and helps to offer a more powerful understanding of the dynamic between these two complexly related cultural practices," from "Science Studies and English Renaissance Literature," *Literature Compass* 3, no. 3 (2006): 341–65. See also Marchitello, *The Machine in the Text: Science and Literature in the Age of Shakespeare and Galileo* (Oxford: Oxford University Press, 2011). Turner and Spiller are also interested in the relationship between practices of reading and writing and the development of the new science.

4. For a useful account of the growing consciousness among historians of science that "presentism or Whiggish historiography" (9) is problematic, see the introduction to Margaret Osler, ed., *Rethinking the Scientific Revolution* (Cambridge: Cambridge University Press, 2000), 3–22.

5. See the introduction to William R. Newman and Anthony Grafton, eds., *Secrets of Nature: Astrology and Alchemy in Early Modern Europe* (Cambridge: MIT Press, 2006), 1–38. See also William R. Newman and Lawrence Principe, *Alchemy Tried in the Fire: Starkey, Boyle, and the Fate of Helmontian Chymistry* (Chicago: University of Chicago Press, 2002).

6. For notable exceptions, see Paula Blank, *Shakespeare and the Mismeasure of Renaissance Man* (Ithaca: Cornell University Press, 2006), who traces Shakespeare's distrust of mathematical measurement and unease with attempts to quantify human values and also Mary Floyd-Wilson, *Occult Knowledge, Science, and Gender on the Shakespearean Stage* (Cambridge: Cambridge University Press, 2013), who traces a more complex interrelationship between old and new beliefs in early modern drama. Literary scholars working on medical texts have been less interested in linking literature to the new science, focusing instead on an exploration of the state of medical knowledge in the sixteenth century and its implications for gender and class. See, e.g., Gail Kern Paster, *The Body Embarrassed: Drama and the Disciplines of Shame in Early Modern England* (Ithaca: Cornell University Press, 1993); Caroline Bicks, *Midwiving Subjects in Shakespeare's England* (Aldershot, UK: Ashgate, 2003); Michael Schoenfeldt, *Bodies and Selves in Early Modern England: Physiology and Inwardness in Spenser, Shakespeare, Herbert, and Milton* (Cambridge: Cambridge University Press, 1999).

7. Reviel Netz, "Counter Culture: Towards a History of Greek Numeracy," *History of Science* 40 (2002): 321.

8. For the importance of embodied perception, see George Lakoff, *Women, Fire, and Dangerous Things: What Categories Reveal about the Mind* (Chicago: University of Chicago Press, 1987). For the role of emotion in rational decision making, see Antonio Damasio, *Descartes' Error: Emotion, Reason, and the Human Brain* (New York: Avon Books, 1994).

9. William Newman, *Atoms and Alchemy: Chymistry and the Experimental Origins of the Scientific Revolution* (Chicago: University of Chicago Press, 2006), 218.

10. There has been a great deal of research on the role of intuitive or naïve science as a barrier to instruction; see, e.g., John Clement, "Students' Preconceptions in Introductory Mechanics," *American Journal of Physics* 50 (1982): 66–71; Andrea diSessa, "Unlearning Aristotelian Physics: A Study of Knowledge-Based Learning," *Cognitive Science* 6 (1982): 37–75; Michael McCloskey, "Naïve Theories in Motion," in Dedre Gentner and A. L. Stevens, eds., *Mental Models* (Hillsdale, NJ: Erlbaum, 1983): 299–324. On the link between cognitive theories of embodiment and intuition, see Paul Cobb, "A Double-Edged Sword: Review of *Intuition in Science and Mathematics* by Efraim Fischbein," *Journal for Research in Mathematics Education* 20 (1989): 214.

11. This definition is derived from Efraim Fischbein, *Intuition in Science and Mathematics: An Educational Approach* (Dordrecht: D. Reidel, 1987), 3–7.

12. See Clement, "Students' Preconceptions." Professor Eric Mazur of Harvard advocates a radical change in the way beginning-level college physics is taught. An article by Craig Lambert in the March–April 2012 issue of *Harvard Magazine* describes how Mazur discovered that at the end of a semester his students "still held the same misconceptions they had at the beginning of the term" because "they basically reverted to an Aristotelian logic, thousands of years back" in order to (mis)understand the relationship between force and weight.

13. Susan Carey, "Cognitive Science and Science Education," *American Psychologist* 41 (1986): 1123.

14. See, however, Ellen Spolsky, *Satisfying Skepticism: Embodied Knowledge in the Early Modern World* (Aldershot, UK: Ashgate, 2001), who argues that from a cognitive perspective, skepticism is a permanent condition of the embodied mind. She focuses on specific manifestations of skepticism in early modern artistic and literary forms that attempt to heal the gap opened up by skeptical doubt, revealing that the embodied mind has a "good enough" access to truth (7). My argument that the cognitive conditions for skepticism were intensified in this period is, I think, compatible with her arguments but has a different emphasis.

15. Joachim Büttner, Peter Damerow, Jürgen Renn, Matthias Schemmel, and Matteo Valleriani, "Galileo and the Shared Knowledge of his Time," Preprint 22B, 2002, Max-Planck-Institut Für Wissenschaftsgeschichte, 10. See also Nancy Nersessian and Lauren B. Resnick, "Comparing Historical and Intuitive Explanations of Motion: Does Naïve Physics Have a Structure?" *Proceedings of the Eleventh Annual Conference of the Cognitive Science Society* (1989): 412–20.

16. Büttner et al., "Galileo," 10.

17. Nancy Nersessian, "Kuhn, Conceptual Change, and Cognitive Science," in Thomas Nickles, ed., *Thomas Kuhn* (Cambridge: Cambridge University Press, 2003), 190–91.

18. Steven Shapin, *The Scientific Revolution* (Chicago: University of Chicago Press, 1996), 53.

19. Dear, *Discipline and Experience*, 22.

20. See Keith Hutchison, "What Happened to Occult Qualities in the Scientific Revolution," in Peter Dear, ed., *The Scientific Enterprise in Early Modern Europe: Readings from Isis* (Chicago: University of Chicago Press, 1997), 93.

21. Peter Dear, "Miracles, Experiments, and the Ordinary Course of Nature," *Isis* 81 (2007): 666.

22. Michel Serres, *The Five Senses: A Philosophy of Mingled Bodies*, trans. Margaret Sankey and Peter Cowley (New York: Continuum, 2008), 22.

23. Gaston Bachelard, *Le Matèrialisme Rationel* (Paris: Presses Universitaires de France, 1971), 41.

24. Literary scholars have tended to emphasize work by historians such as Shapin and Shaffer who focus on the cultural embeddedness of the new science. Newman, *Atoms and Alchemy*, 8–11, has criticized these historians for retaining an outdated account of the intellectual history of the period, arguing that "contextualist" historians of science like Shapin and Schaffer have essentially added "sociological explanations to the preexisting history of ideas rather than subjecting the results of intellectual history to critical analysis" (9).

25. See Newman, *Atoms and Alchemy*, Hutchison, "What Happened," and Shapin, *Scientific Revolution*, and also essays by Margaret Osler ("Galileo, Motion, and Essences," 107–12) and Christoph Meinel ("Early Seventeenth-Century Atomism: Theory, Epistemology, and the Insufficiency of Experiment," 176–211) in Dear, *The Scientific Enterprise*. Hutchison argues that the new science did not, as has often been argued, reject "occult qualities" but instead began to accept only "occult" causes for natural phenomena. Osler outlines the "essentialism" of the Aristotelian position that "our ob-

servations in some manner reveal the inner natures of objects to us" (107), while later thinkers like Galileo, Gassendi, and Locke understood that "empirical knowledge must in principle be confined to appearances and probabilities" (112). Meinel argues that "there was no experimental proof possible" until the nineteenth century for the atomic theory of matter advocated by seventeenth-century scientists like Robert Boyle (176).

26. See Craig Martin, *Renaissance Meteorology: Pomponazzi to Descartes* (Baltimore: Johns Hopkins University Press, 2011), 15, who points to William Newman, *Atoms and Alchemy*; Roger Ariew, *Descartes and the Last Scholastics* (Ithaca: Cornell University Press, 1999); and Margaret Osler, "New Wine in Old Bottles: Gassendi and the Aristotelian Origin of Early Modern Physics," *Midwest Journal of Philosophy* 26 (2002): 151–68. These scholars do offer different "Aristotles." Martin argues for the centrality of the *Meteorology*, whereas Osler, whose article is based on the work of the French atomist Gassendi, argues that Aristotle's *Physics* offered an influential theory of matter. I agree with Martin and with Eckhard Kessler, "Metaphysics of Empirical Science," in Marianne Pade, ed., *Renaissance Readings of the Corpus Aristotelicum* (Copenhagen: Museum Tusculanum Press, 2001), 79–80, that Aristotle's more metaphysical works remained more important in France than in Italy or England.

27. Deborah E. Harkness, *John Dee's Conversations with Angels: Cabala, Alchemy, and the End of Nature* (Cambridge: Cambridge University Press, 1999), argues that the angel conversations should be seen in the context of Dee's interest in "the possibilities of a new optical, experimentally based natural philosophy." She notes that Dee "did not rely on conjurations or binding spells, but on simple prayer and the use of a crystal stone to collect the occult rays which all major theological and philosophical authorities agreed were vehicles for angelic, as well as astral, influences" (96). She concludes that "Dee would not have considered his angel conversations a 'magical' practice" but rather involved an attempt "to unify and make coherent all religious beliefs, natural knowledge, and ancient theory" (130).

28. Thomas Kuhn, *The Road Since Structure: Philosophical Essays, 1970–1993, with an Autobiographical Interview*, ed. James Conant and John Haugeland (Chicago: University of Chicago Press, 2000), 15–17.

29. Ibid., 16–17.

30. See Paul Hoyingen-Hume, *Reconstructing Scientific Revolutions: Thomas Kuhn's Philosophy of Science*, trans. Alexander T. Levine (Chicago: University of Chicago Press, 1993), 131–35, on the problems with the term and Kuhn's changing definition of it. For an early critique of the concept, see Margaret Masterman, "The Nature of a Paradigm," Imre Lakotos and Alan Musgrave, eds., *Criticism and the Growth of Knowledge* (Cambridge: Cambridge University Press, 1970), who argues that in *The Structure of Scientific Revolutions*, Kuhn uses the term "paradigm" in "not less than twenty-one different senses" (61).

31. Nersessian, "Kuhn," 184–85.

32. Kuhn, "Commensurability, Comparability, Communicability," in *The Road Since Structure*, 34.

33. Translation from Gary Gutting, *Michel Foucault's Archaeology of Scientific Reason* (Cambridge: Cambridge University Press, 1989), 14. "Nous croyons, en effet, que le progress scientifique manifeste toujours une rupture, de perpétuelles ruptures, entre conaissance commune et conaissance scientifique," Bachelard, *Le Matérialisme Rationel*, 207.

34. Nersessian, "Kuhn," notes that Kuhn's ideas about "concept representation and acquisition" anticipated the work of Eleanor Rosch on categorization (181).

35. By *Aristotelian naturalism,* I mean something quite different from the phrase *Renaissance naturalism,* sometimes used to refer to Hermeticism or other so-called occult bodies of knowledge.

36. Charles Schmitt, *John Case and Aristotelianism in Renaissance England* (Oxford: Oxford University Press, 1983). See, e.g., William West, "What's the Matter with Shakespeare?: Physics, Identity, Playing," *South Central Review* 26 (2009): 103–26, who bases his account of Aristotelian matter theory on the *Physics, Metaphysics,* and *Categories.* To the extent that he focuses on controversy about transubstantiation, where concepts from the *Physics* were routinely deployed, his account is correct. But I think he errs in assuming that these works comprised Aristotelian matter theory in the period and that hylomorphism was its dominant feature.

37. Stephen Toulmin and June Goodfield, *The Architecture of Matter* (Chicago: University of Chicago Press, 1962), 142. They emphasize the range of theories about matter available at the beginning of the seventeenth century, including Aristotelian ideas about *minima,* the ideas of Plato, pneuma-theory of the Stoics, and Lucretian atomism (143).

38. Newman, *Atoms and Alchemy,* 218.

39. Hutchison, "What Happened," 98.

40. Shapin, *Scientific Revolution,* 55.

41. I'm using *naturalism* in the sense that Kessler uses it and not in the sense employed by Shapin in *Scientific Revolution,* 43. Toulmin and Goodfield, in *Architecture of Matter,* use the phrase "Aristotle's functional matter-theory" to describe this set of concepts (85).

42. Newman, *Atoms and Alchemy,* 13. Hilary Gatti, "The Natural Philosophy of Thomas Harriot," in Robert Fox, ed., *Thomas Harriot: An Elizabethan Man of Science* (Aldershot, UK: Ashgate, 2000), has shown how Harriot based his natural philosophy in part on "a naturalistic interpretation of Aristotle," derived from an Italian commentary on the *Meteorology* and the work of Pietro Pomponazzi (70).

43. Martin, *Renaissance Meteorology,* 22, 19, 15, 16.

44. I use the term *secret* rather than *occult* to avoid the confusion associated with the modern definition of occult as involving supernatural agency. A number of historians of science have debunked the views of Frances Yates and Muriel Bradbrook that identified a tradition of occultism based in Hermeticism and have argued instead that in England thinkers like Dee and Harriot combined elements of Aristotelian naturalism with ideas derived from the works of Roger Bacon and from alchemical texts. On the "secrets of nature" tradition, see William Eamon, *Science and the Secrets of Nature: Books of Secrets in Medieval and Early Modern Culture* (Princeton: Princeton University Press, 1994); Newman and Grafton, *Secrets of Nature*; Newman, *Atoms and Alchemy.* On the problems with the Yates thesis and the influence of Aristotelian naturalism and the works of Roger Bacon on Thomas Harriot, see Stephen Clucas, "Thomas Harriot and the Field of Knowledge in the English Renaissance," in Fox, *Thomas Harriot,* 93–136. On the influence of naturalism and Bacon on John Dee, see Nicholas Clulee, "Astrology, Magic and Optics: Facets of John Dee's Early Natural Philosophy," *Renaissance Quarterly* 30 (1977): 632–80.

45. Stephen Greenblatt, *The Swerve: How the World Became Modern* (New York: Norton, 2011).

46. For Lucretian influence on Shakespeare and Marlowe, see R. Allen Shoaf, "'If Imagination Amend Them': Lucretius, Marlowe, Shakespeare," in Gordon Bradshaw, Tom Bishop, and David Schalwyk, eds., *The Shakespearean International Yearbook* 4, *Shakespeare Studies Today* (Aldershot, UK: Ashgate, 2010), 257–80, who argues that Shakespeare's *A Midsummer Night's Dream* is arguing against Marlowe's Lucretianism. I'm not fully convinced by the echoes of Lucretius he finds in Shakespeare and Marlowe. See also Jonathan Gil Harris, "Atomic Shakespeare," *Shakespeare Studies* 30 (2002): 47–51, who traces the appearance of the word *atom* in Shakespeare's works. Harris notes that Shakespeare may have known about the *De Rerum Natura* through Montaigne or the works of Girolamo Fracastoro but acknowledges that there is no direct evidence of Lucretian influence on Shakespeare, although he argues that "Lucretian philosophy resonates with a noteworthy materialist strand in his plays" (48). David Riggs, *The World of Christopher Marlowe* (New York: Henry Holt, 2004), 88–89, argues unconvincingly that Marlowe derived a belief in a "materialistic, ever-changing cosmos" from Ovid's *Metamorphoses*. Kristen Poole, *Supernatural Environments in Shakespeare's England: Spaces of Demonism, Divinity, and Drama* (Cambridge: Cambridge University Press, 2011), rightly argues that New Historicist critics have not taken early modern belief in the supernatural seriously enough. However, her argument that there was an identifiable "Ovidian physics" characterized by belief in a world of flux or change (51–57) may not adequately consider the theories of change offered by Aristotelian sublunar physics, especially with regard to the imperfect mixtures of meteorology. David A. Hedrich Hirsch, "Donne's Atomies and Anatomies: Deconstructed Bodies and the Resurrection of Atomic Theory," *SEL* 31 (1991): 69–94, cites G. B. Stones, "The Atomic View of Matter in the XVth, XVIth, and XVIIth Centuries," *Isis* 10 (1928): 445–65, as support for the ancient sources of Donne's interest in atomism. Understanding of the sources of seventeenth-century atomism has changed a great deal since 1928.

47. See Newman, *Atoms and Alchemy*, who argues that Boyle's atomism was more influenced by alchemical naturalism than by ancient atomism. See also Toulmin and Goodfield, *Architecture of Matter*, 143. See also Antonio Clericuzio, *Elements, Principles, and Corpuscles: A Study of Atomism and Chemistry in the Seventeenth Century* (Dordrecht: Kluwer, 2010), 3–5. Richard Westfall, *The Construction of Modern Science: Mechanisms and Mechanics* (Cambridge: Cambridge University Press, 1977), 65–81, emphasizes the role of experimental observation in the development of mechanical chemistry in the seventeenth century. J. A. Bennett, "The Mechanics' Philosophy and the Mechanical Philosophy," *History of Science* 24 (1986): 1–28, argues that "literary references aside, ancient atomism would have been a very weak and ineffectual model for the mechanical philosophy" (4).

48. Clericuzio points to French influence on the members of the "Newcastle Circle," whose atomism was directly inspired by Lucretius (86–88).

49. Newman, *Atoms and Alchemy*, 225, argues that "Boyle and his peers" were like "a conquering horde bent on executing its prisoners rather than interrogating them," allowing "only a distorted picture of the previous order to remain."

50. Dear, *Discipline and Experience*, 24.

51. See William M. Hamlin, *Tragedy and Scepticism in Shakespeare's England* (Houndmills, Basingstoke: Palgrave, 2005), 29–71, who provides a comprehensive ac-

count of possible traces of skeptical thinking in Elizabethan England. He identifies skeptical ideas in traditional proverbs and shows that there was some access to ancient and continental philosophical skepticism, but he has a hard time demonstrating clear influence of ancient skepticism on English writers before 1603.

52. Ibid., 8.

53. Spolsky, *Satisfying Skepticism,* 10. As noted above, her focus is on early modern artifacts that work to reassure that the brain's knowledge of the world is "good enough."

54. Richard Popkin, *The History of Scepticism From Savonarola to Boyle,* rev. ed. (Oxford: Oxford University Press, 2003).

55. Clulee, "Astrology," 633–35, argues convincingly that Dee's early works were informed by a number of sources including Aristotelian naturalism and the optics of medieval thinkers like Roger Bacon.

56. See Marchitello, *Machine,* 12–15, who argues that "another goal of our contemporary studies is the telling of new narratives about the unity of culture-science that has always been in place but has always been obscured by the ideological separation of science from literature, and the history of construction of the separation between science and the rest of culture" (12). Carla Mazzio reminds us that in the sixteenth century, "science *was* knowledge" and comprised "the quadrivium (arithmetic, geometry, astronomy, and music) *and* the trivium (rhetoric, grammar, and logic)" (1–2).

57. Newman and Grafton, *Secrets of Nature,* 17–18. They refer to the final chapters of Keith Thomas, *Religion and the Decline of Magic* (New York: Scribners, 1971), 631–32; Brian Vickers, "On the Function of Analogy in the Occult," in A. G. Debus and Ingrid Merkels, eds., *Hermeticism in the Renaissance* (Washington: Folger Books, 1988), 265–92; Frances Yates, *The Rosicrucian Enlightenment* (London: Routledge and Kegan Paul, 1972), ix–xii.

58. Newman and Grafton, *Secrets of Nature,* 15. I do find it useful to group together a range of fields, including alchemy and astrology, as focusing on hidden and secret rather than manifest and intuitive knowledge. But I acknowledge that they work out their relation to hidden knowledge in very different ways, based on very different assumptions.

59. See Stuart Clark's excellent *Thinking with Demons: The Idea of Witchcraft in Early Modern Europe* (Oxford: Oxford University Press, 1997), 211–13, on Reginald Scott in this Protestant naturalistic context. See also Richard Bauckham, "Science and Religion in the Writings of Dr William Fulke," *British Journal for the History of Science* 8 (1975): 17–31, for Fulke's extreme naturalism and its Protestant context.

60. See William Newman's clear critique of Brian Vickers's problematic definition of the early modern "occult," in "Brian Vickers on Alchemy and the Occult: A Response," *Perspectives on Science* 17 (2009), 485–88; see also Clark, *Thinking with Demons,* 215–20.

61. As Newman has argued, the introduction to Brian Vickers, ed., *Occult and Scientific Mentalities in the Renaissance* (Cambridge: Cambridge University Press, 1984), 1–56, is an example of this approach. Many of the essays in the volume take a more historically nuanced view of the occult, however.

62. Vickers is again an example of this kind of thinking. Michel Foucault, *The Order of Things: An Archaeology of the Human Sciences* (New York: Vintage, 1970), exaggerates the centrality of resemblance in premodern thought. Many scholars follow Foucault in

attributing too great a role to correspondences in the sixteenth century, see, e.g., Jonathan Sawday's otherwise very useful *The Body Emblazoned: Dissection and the Human Body in Renaissance Culture* (London: Routledge, 1996), which refers to the "endlessly repetitive interplay of metaphor, similitude, and comparison" within which "the body lay entangled within a web of enclosing patterns of repetition" (23). Some writers in the period do represent the body in this way, but it was not quite so universal or inescapable a discourse. See also Katherine Eggert, "*The Alchemist* and Science," in Garrett A. Sullivan, Jr., Patrick Cheney, and Andrew Hadfield, eds., *Early Modern English Drama: A Critical Companion* (Oxford: Oxford University Press, 2006), 201, for a clear statement of the partial truth that "European scientific knowledge before the seventeenth century understood the world through the resemblances of things." Some writers did share this understanding, but it did not constitute "European scientific knowledge before the seventeenth century," as I hope this book will show. Eggert's sources for this account of premodern science explain why she takes this view. They are A. O. Lovejoy, *The Great Chain of Being*, 2nd ed. (New York: Harper and Row, 1960); Foucault, *Order of Things*; Eamon, *Science and the Secrets of Nature*; Frances Yates, *The Occult Philosophy in the Elizabethan Age*. Eggert makes no reference to more recent work that has questioned this account, such as Newman and Grafton, *Secrets of Nature*, or the work of Clark, Clucas, and Clulee.

63. See works by Frances Yates, including *Giordano Bruno and the Hermetic Tradition* (Chicago: University of Chicago Press, 1964) and *The Rosicrucian Enlightenment*. The "Yates thesis" has been widely questioned and critiqued by scholars as different as Brian Vickers, *Occult and Scientific Mentalities*, 5–6, and Stuart Clark, *Thinking with Demons*, 227, who argues that "consideration of things occult" did not "commit natural philosophers and academic physicians to what Frances Yates called 'the occult philosophy.'"

64. Newman and Grafton, *Secrets of Nature*, 5, argue that astrology should be regarded as "one of the many highly practical sets of intellectual tools that Renaissance thinkers forged and honed for dealing with the same problems that they also attacked with what now seem the shinier tools of social and political analysis." See also Newman's critique of Vickers's incorrect assumptions about the irrational nature of the occult in "Brian Vickers on Alchemy," 482–506.

65. Floyd-Wilson, *Occult Knowledge*, 2.

66. A number of scholars have traced early modern attitudes toward the environment and the interdependence of humans and the world. See, esp., Lynne Bruckner and Dan Brayton, eds., *Ecocritical Shakespeare* (Aldershot, UK: Ashgate, 2011); Robert N. Watson, *Back to Nature: The Green and the Real in the Late Renaissance* (Philadelphia: University of Pennsylvania Press, 2006): Jeffrey Theis, *Writing the Forest in Early Modern England: A Sylvan Pastoral Nation* (Pittsburgh: Duquesne University Press, 2010).

67. Brian Greene, *The Hidden Reality: Parallel Universes and the Deep Laws of the Cosmos* (New York: Vintage, 2011), vii.

68. Lewis Wolpert, *The Unnatural Nature of Science* (Cambridge: Harvard University Press, 1992), xi.

69. Bruno Latour, *An Inquiry into Modes of Existence: An Anthropology of the Moderns*, trans. Catherine Porter (Cambridge: Harvard University Press, 2013), 3.

70. Wolpert, *Unnatural Nature*, x, 177.

CHAPTER 2: Aristotelian Naturalism and Its Discontents

1. Stevin Shapin, *The Scientific Revolution* (Chicago: University of Chicago Press, 1996), in his first two chapters (15–118), considers "What was known?" and "How was it known?" in Europe at the cusp of the scientific revolution. He is concerned to show how ideas changed all over Europe. In contrast, I'm interested in a more fine-grained way in what educated people in England knew, and how they thought they knew it, in the eighty-five years before the publication of Galileo's *Sidereus Nuncius* in 1610.

2. Shapin, *Scientific Revolution*, notes that this account of what came before the scientific revolution is exaggerated because it is derived from the claims of seventeenth-century proponents of the "new science," who were concerned to emphasize the novelty of their own arguments: "corpuscular and mechanical philosophers . . . vigorously insisted that their innovations represented radical departures from traditionally constituted bodies of knowledge" (65). Shapin goes on to argue that "historians and philosophers of science have traditionally paid far too much attention to formal methodological pronouncements, often taking such statements as adequate accounts of what past practitioners actually did" (94–95).

3. See Charles B. Schmitt, *John Case and Aristotelianism in Renaissance England* (Kingston: McGill-Queens University Press, 1983), 23, for the idea that "late fifteenth- and sixteenth-century European Aristotelianism was by no means uniform and monolithic. Rather, I think it best to speak of 'Aristotelianisms' to denote the fact."

4. See Stephen Toulmin and June Goodfield, *The Architecture of Matter* (Chicago: University of Chicago Press, 1962), 142, who argue about alchemy that "despite the enveloping mass of trash, a core of sound doctrine and practical recipes got through: serious workers (such as Newton) could tell one from the other as readily as a modern scientist can distinguish pulp science-fiction from the *Proceedings of the Royal Society*." See also William Newman, "Brian Vickers on Alchemy and the Occult: A Response," *Perspectives on Science* 17 (2006): 484–506, for a refutation of Vickers's arguments in *Occult and Scientific Mentalities in the Renaissance* (Cambridge: Cambridge University Press, 1984), that occult knowledge and practices did not contribute to the development of science. Stuart Clark, "The Scientific Status of Demonology," an essay found in Vickers's volume, argues convincingly that treatises on demonology and witchcraft often provide complicated analysis of the status of demonic activity: "it must be stressed, therefore, that demonic intervention did not turn natural into supernatural causation. . . . The distinguishing criterion of demonic, and indeed all forms of magic, was not that it was supernatural but that it was unusual" (362).

5. See Schmitt, *John Case*, 28, who argues that "Aristotelianism" in late sixteenth-century England often included "Neoplatonic, Hermetic, Stoic, or alchemical" elements. Craig Martin, *Renaissance Meteorology: Pomponazzi to Descartes* (Baltimore: Johns Hopkins University Press, 2011), similarly argues that Aristotle's *Meteorology* continued to influence "anti-Aristotelians" such as Descartes (149).

6. Peter Dear, "Miracles, Experiments, and the Ordinary Course of Nature," *Isis* 81 (1990): 665, contrasts an Aristotelian view that knowledge about the "ordinary course of nature" was "common property" with traditions that saw knowledge as secret. See also William Eamon, *Science and the Secrets of Nature: Books of Secrets in Medieval and Early Modern Culture* (Princeton: Princeton University Press, 1994), 22–23, for the dis-

tinction between the Aristotelian focus on "the regularities in nature" and a Hellenistic belief that "the really essential features of nature were hidden from the senses as well as from the intellect" (23).

7. See Peter Dear, "Miracles, Experiments," 665–66, for the Aristotelian belief that the truth about nature comes from universal, everyday experience and not from witnessing a "singular event" (666).

8. Eamon, *Science and the Secrets of Nature*, 54.

9. Peter Dear, *Discipline and Experience: The Mathematical Way in the Scientific Revolution* (Chicago: University of Chicago Press, 1995), explains the difference between Aristotelian "experience" and the new concept of "experiment" in the seventeenth century (22–23).

10. Lorraine Daston, "Preternatural Philosophy," in Daston, ed., *The Biographies of Scientific Objects* (Chicago: University of Chicago Press, 2000): 15–41, charts the increased emphasis on the study of "preternatural" (as opposed to natural and supernatural) phenomena in the late sixteenth and early seventeenth centuries, and its ultimate decline in the late seventeenth and early eighteenth centuries. These phenomena are "secret" in the sense that they are hidden and "preternatural" in the sense that they are unusual. I'm using "secret" because I'm more interested in their epistemological implications.

11. James Bono, *The Word of God and the Languages of Man* (Madison: University of Wisconsin Press, 1995), 34.

12. Bono, *The Word*, 35.

13. On this "secrets" tradition, see Eamon, *Science and the Secrets of Nature*. The role of "natural magic" and "Hermeticism" in the scientific revolution has been controversial ever since Frances Yates, who in works like *Giordano Bruno and the Hermetic Tradition* (London: 1964; reprint, University of Chicago Press, 1991) argued for its important role in the scientific revolution. For a recent account of the secrets tradition in England, see Brian Copenhaver, "Natural Magic, Hermeticism, and Occultism in Early Modern Science," in David C. Lindberg and Robert. S. Westman, eds., *Reappraisals of the Scientific Revolution* (Cambridge: Cambridge University Press, 1990), 261–302.

14. Daston, "Preternatural Philosophy," 27.

15. Newman, in his "Brian Vickers on Alchemy," is critical of Vickers's assumption that practices like alchemy, astrology, and Cabala can be lumped together as "occult." Like Newman, I reject Vickers's definition of "occult" as involving superstition and understand that the practices listed above were different in many respects. But they do all share a focus on phenomena that were, in the Aristotelian system, not manifest, and late sixteenth-century English writers like Dee do tend to combine aspects of several such practices.

16. Schmitt, *John Case*, especially "Aristotelianism in England," (3–12).

17. See Frank A. Lewis and Robert Bolton, *Form, Matter, and Mixture in Aristotle* (Oxford: Blackwell, 1996), for a series of chapters articulating some of the problems and inconsistencies in Aristotle's theory of matter.

18. Eckhard Kessler, "Metaphysics or Empirical Science?" in Marianne Pade, ed., *Renaissance Readings of the Corpus Aristotelicum* (Copenhagen: Museum Tusculanum Press, 2001), 79–80.

19. See William R. Newman, *Atoms and Alchemy: Chymistry and the Experimental Origins of Modern Science* (Chicago: University of Chicago Press, 2006), for Aristotelian naturalism in alchemy. For the role of naturalism in the thought of John Dee, see Nicholas Clulee, "Astrology, Magic, and Optics: Facets of John Dee's Early Writings," *Renaissance Quarterly* 30 (1977): 632–80; on naturalism in Harriot, see Stephen Clucas, "Thomas Harriot and the Field of Knowledge in the English Renaissance," in Robert Fox, ed., *Thomas Harriot, an Elizabethan Man of Science* (Aldershot, UK: Ashgate, 2000), 93–136. On naturalism and Aristotle's *Meteorology*, see Martin, *Renaissance Meteorology*.

20. Kessler, "Metaphysics," 80. See Robert Bolton, "Science and the Science of Aristotle," in Lewis and Bolton, *Form, Matter*, 232–34, on whether the discussion in the *Metaphysics* needs to be reconciled with the treatments of matter in his other works; in the same book, see also Kit Fine, "The Problem of Mixture," for a thorough treatment of the problem of explaining mixture in terms of the elements while preserving hylomorphic theory (82–182).

21. On Boyle's critique of substantial forms, see Peter Dear, "Intelligibility in Science," *Configurations* 11 (2003): 152. On the development of this concept in the Middle Ages, see Copenhaver, "Natural Magic," 272–73.

22. Kessler, "Metaphysics," 86.

23. Ibid., 89–90.

24. Martin, *Renaissance Meteorology*, 15–16.

25. See Francis R. Johnson, *Astronomical Thought in Renaissance England* (Baltimore: Johns Hopkins University Press, 1937), 68–73. These treatises are based on Aristotle, Ptolemy, and Arabic commentators on those texts.

26. Ann Blair, "The *Problemata* as a Natural Philosophical Genre," in *Natural Particulars: Nature and the Disciplines in Renaissance Europe*, ed. Anthony Grafton and Nancy Siraisi (Cambridge: MIT Press, 1999), 171.

27. Ibid., 185.

28. Ibid., 185.

29. Ibid., 173.

30. Ibid., 174.

31. Eamon, *Science and the Secrets of Nature*, 54.

32. G. C. Moore Smith, ed., *Gabriel Harvey's Marginalia* (Stratford-Upon-Avon: Shakespeare Head Press, 1913, 163.

33. Eamon, *Science and the Secrets of Nature*, 45.

34. Ibid., 51–53.

35. Ibid., 292–93.

36. On the essentialism of Aristotelian theories of nature, see Margaret Osler, "Galileo, Motion, and Essences," in Peter Dear, ed., *The Scientific Enterprise in Early Modern Europe: Readings from Isis* (Chicago: University of Chicago Press, 1997), 108, for the Aristotelian position that "from the observable characteristics of substances we can learn something of their inner essences."

37. Shapin, *Scientific Revolution*, 28–29, emphasizes that the new mechanistic atomism of the seventeenth century "assaulted common sense," since the substrate of atoms couldn't be seen, unlike the Aristotelian elements. He argues that the teleological nature

of Aristotle's theory of motion coincided with the human tendency to offer "teleological—or goal-oriented—accounts of their own movements." So the movement of objects makes sense by analogy with intentional human movement (29–30).

38. John Rastell, *Four Elements,* in Richard Axton, ed., *Three Rastell Plays* (Cambridge: D. S. Brewer, 1979), 33.

39. Kuhn, "What are Scientific Revolutions?" in Lorenz Kruger, Lorraine J. Daston, and Michael Heidelberger, eds., *The Probabilistic Revolution,* vol. 1 of *Ideas in History* (Cambridge: MIT Press, 1987), 10.

40. Kuhn, "What are Scientific Revolutions?" 10.

41. Shapin, *Scientific Revolution,* 29.

42. Fine, "The Problem," 85, explains the problem of mixtures, and his explanation makes clear that Rastell's account leaves out many details. Fine notes that an atomic theory of matter does solve most of the problems of mixture in a way that Aristotelian matter theory cannot (90).

43. Newman, "Brian Vickers on Alchemy," notes that *occult* was used in the early modern period in two senses: to denote "occult qualities" like magnetism and to describe knowledge that needed to be kept hidden (485–86). Newman argues that neither usage corresponds to a modern sense of *occult* as supernatural. See also Keith Hutchison, "What Happened to Occult Qualities in the Scientific Revolution?" in Peter Dear, ed., *The Scientific Enterprise,* 87.

44. Eamon, *Science and the Secrets of Nature,* 54.

45. See Hutchison, "What Happened," 87–95. For an account of occult forces and their special relation to issues of gender, see Mary Floyd-Wilson, *Occult Knowledge, Science, and Gender on the Shakespearean Stage* (Cambridge: Cambridge University Press, 2013).

46. Eamon, *Science and the Secrets of Nature,* 285–300, argues that the "new philosophy" involved a hunt for natural causes of occult phenomena. Hutchison, "What Happened," argues that the rejection of "occult qualities" in the seventeenth century was really a rejection of the idea that any causal phenomena were manifest. Lorraine Daston and Katherine Park, *Wonders and the Order of Nature, 1150–1750* (New York: Zone Books, 2001), discuss the focus of "preternatural philosophers" on occult phenomena (160–67).

47. See Newman, "Brian Vickers on Alchemy," 486, who argues that astrology and alchemy "were often thought to work by means of the Aristotelian primary qualities," and Eamon, *Science and the Secrets of Nature,* 194, who argues that natural magic was "the science that attempted to give rational, naturalistic explanations of the occult forces of nature."

48. S. K. Heninger, Jr., *A Handbook of Renaissance Meteorology with Particular Reference to Elizabethan and Jacobean Literature* (Durham, NC: Duke University Press, 1960), 3–4. See also Martin, *Renaissance Meteorology,* 1–20.

49. Heninger, *A Handbook,* 9.

50. Ibid., 9.

51. See Craig Martin, "The Ends of Weather: Teleology in Renaissance Meteorology," *Journal of the History of Philosophy* 48 (2010): 259–82, for a discussion of the difference between Italian discussions of the *Meteorology,* which avoid speculation about teleology and stick to naturalistic explanations, and Lutheran writings on this work, which

argue that God as "final cause" uses meteorological events as warnings or for other purposes.

52. Leonard Digges, A *prognostication everlasting of right good effect* . . . (London, 1592), EEBO, accessed April 11, 2011. I'm citing the later edition, which includes Thomas Digges's addendum.

53. Johnson, *Astronomical Thought,* 16–65, still provides a useful account of pre-Copernican astronomy.

54. Thomas Kuhn, *The Copernican Revolution* (Cambridge: Harvard University Press, 1957), 47–50, on "the problem of the planets."

55. Ibid., 59–64.

56. Ibid., 73.

57. Ibid., 268–69.

58. On the distinction between natural and judicial astrology in the work of John Dee, see Richard Dunn, "John Dee and Astrology in Elizabethan England," in Stephen Clucas, ed., *John Dee: Interdisciplinary Studies in Renaissance Thought* (Dordrecht: Springer, 2010), 85–96.

59. J. L. Heilbron, introduction to *John Dee on Astronomy: Propaedumata Aphoristica (1558 & 1568),* ed. and trans. Wayne Shumaker (Berkeley and Los Angeles: University of California Press, 1978), 50–51, 67–73.

60. Deborah Harkness, *John Dee's Conversations with Angels: Cabala, Alchemy, and the End of Nature* (Cambridge: Cambridge University Press, 1999), 118.

61. Ibid., 117. Harkness cites Clulee, "Astrology, Magic, and Optics," 678, for establishing the importance of astral radiation in Dee's natural philosophy. However, she goes farther than Clulee in seeing even the angel conversations as having a basis in natural philosophy.

62. Roy Porter, *Disease, Medicine, and Society in England, 1550–1860,* 2nd ed. (Cambridge: Cambridge University Press, 1995), 12–13.

63. On the influence of Aristotle on Galen, see Nancy Siraisi, *Medieval and Early Renaissance Medicine: An Introduction to Knowledge and Practice* (Chicago: University of Chicago Press, 1990), 3. From antiquity, medicine also developed its own separate "secrets" traditions involving occult forces, for which see Floyd-Wilson, *Occult Knowledge,* 6–20.

64. Gail Kern Paster, *The Body Embarrassed: Drama and the Disciplines of Shame in Early Modern England* (Ithaca: Cornell University Press, 1993), 7.

65. Siraisi, *Medieval,* 102–3.

66. Mary Lindemann, *Medicine and Society in Early Modern Europe* (Cambridge: Cambridge University Press, 2010), notes that the "new" Vesalian anatomy was largely based "on the precepts of Galen" (91), and that Galenic medicine remained dominant "at least until the middle of the seventeenth century" (87). I disagree with the argument of Jonathan Sawday, *The Body Emblazoned: Dissection and the Human Body in Renaissance Culture* (New York: Routledge, 1995), to the extent that he follows Foucault in positing a dominant premodern understanding of the body as "entangled within a web of enclosing patterns of repetition" (23).

67. Siraisi, *Medieval,* 129.

68. See Vivian Nutton, "The Seeds of Disease: An Explanation of Contagion and Infection from the Greeks to the Renaissance," *Medical History* 27, no. 1 (1983): 1–34.

She argues that in Galenic medicine, "contagion was discussed in terms of what could be observed, but the temptation was always there to seek for its invisible causes" (2). Linda Deer Richardson, "The Generation of Disease: Occult Causes and Diseases of the Total Substance," in A. Wear, R. K. French, and M. Lonie, eds., *The Medical Renaissance of the Sixteenth Century* (Cambridge: Cambridge University Press, 1985), 175–94, traces the controversial theories of French medical writer Jean Fernel, whose *De Abditis Rerum Causis* [Concerning the Hidden Causes of Things] argues that there are non-humoral hidden causes for contagious disease but dismisses magical cures that involve charms.

69. Geoffrey Eatough, ed., *Fracastoro's Syphilis* (Liverpool: Francis Cairns, 1984). The first line of the poem refers to "semina morbum" or "seeds of disease."

70. Siraisi, *Medieval,* 193.

71. Walter Pagel, *Paracelsus: An Introduction to Philosophical Medicine in the Era of the Renaissance,* 2nd rev. ed. (Basel: S. Karger, 1982), 131–32.

72. R. B. (usually identified as Richard Bostocke), *The Difference betweene the Auncient Phisicke . . . and the latter Phisicke* (London, 1585), C6v.

73. Pagel, *Paracelsus,* 140–41.

74. Bono, *The Word,* 134.

75. Ibid., 135.

76. Pagel, *Paracelsus,* 146, 148.

77. Sawday, *The Body Emblazoned,* 232.

78. Michel Foucault, *The Order of Things: An Archaeology of the Human Sciences* (New York: Vintage Books, 1973). The footnotes to chapter 2, laying out his account of the "four similitudes" that characterized understanding of nature before the scientific revolution (45), reveal the extent to which it is based in the writings of Paracelsus. But see also Dedre Gentner and Michael Jeziorski, "The Shift from Metaphor to Analogy in Western Science," in A. Ortony, ed., *Metaphor and Thought,* 2nd ed. (Cambridge: Cambridge University Press, 2002), 447–80, who take alchemical signatures as representative of pre–scientific revolution thought. Marjorie Hope Nicolson, *The Breaking of the Circle: Studies in the Effect of the New Science Upon Seventeenth Century Poetry* (Evanston, IL: Northwestern University Press, 1950), also relies heavily on Paracelsus for images of microcosm, macrocosm, correspondence, and the unbroken circle (11, 12–13, 15, 99, 109, 130).

79. See Markus Asper, "The Two Cultures of Mathematics in Ancient Greece," in Eleanor Robson and Jacqueline Stedall, eds., *The Oxford Handbook of the History of Mathematics* (Oxford: Oxford University Press, 2011), 107–32, and Reviel Netz, "Counter Culture: Towards a History of Greek Numeracy," *History of Science* 40 (2002): 321–52.

80. Morris Kline, *Mathematical Thought from Ancient to Modern Times* (New York: Oxford University Press, 1972), 43.

81. Ibid., 153.

82. Alfred W. Crosby, *The Measure of Reality: Quantification and Western Society, 1250–1600* (Cambridge: Cambridge University Press, 1997), 13, 47.

83. Stephen Chrysomalis, "The Cognitive and Cultural Foundations of Numbers," in Robson and Stedall, *Oxford Handbook,* 504–5.

84. Alfred W. Crosby, *The Measure of Reality: Quantification and Western Society, 1250–1600* (Cambridge: Cambridge University Press, 1997), 43.

85. Netz, "Counter Culture," 324.

86. Richard W. Hadden, *On the Shoulders of Merchants: Exchange and the Mathematical Conception of Nature in Early Modern Europe* (Albany: SUNY Press, 1994), 67. For the essentialist nature of ancient mathematics and the development of more abstract mathematics during the scientific revolution, see Jacob Klein, *Greek Mathematical Thought and the Origin of Algebra* (Cambridge, MA: MIT Press, 1968; reprint, New York: Dover, 1992), 122–25.

87. Robert Recorde, *The Whetstone of Witte* (London, 1557), Early English Books Online (EEBO), accessed April 11, 2011, A2r, A2v.

88. Kline, *Mathematical Thought*, 173.

89. Ibid., 75.

90. Ibid., 143

91. Ibid., 202.

92. For bookkeeping schools and basic mathematics books in sixteenth-century England, see Linda Woodbridge, introduction to *Money and the Age of Shakespeare: Essays in New Economic Criticism*, ed. Linda Woodbridge (New York: Palgrave Macmillan, 2003), 1–18.

93. Frank J. Swetz, *Capitalism and Arithmetic: The New Math of the 15th Century* (La Salle, IL: Open Court Publishing, 1987; second printing, 1989), 27. See also Gillian Evans, "From Abacus to Algorism: Theory and Practice in Medieval Arithmetic," *British Journal for the History of Science* 10 (1977): 114–31, who traces the split between new practices of "algorism" and the number theory of antiquity, transmitted by Boethius, that was incompatible with it.

94. Hadden, *On the Shoulders*, 63.

95. Dorothy Schrader, "The Arithmetic of the Medieval Universities, in Swetz, *Capitalism*, 23.

96. Crosby, *Measure*, 119.

97. Hadden, *On the Shoulders*, 90–95.

98. Johnson, *Astronomical Thought*, 95, makes this point.

99. See Jacqueline Stedall, "Tracing Mathematical Networks in Seventeenth-Century England," in Robson and Stedall, ed., *Oxford Handbook*, 133–52.

100. Netz, "Counter Culture," 345–46, associates the spread of Hindu-Arabic numerals with the development of modern capitalism and modern science.

101. For Gabriel Harvey's lists of contemporary mathematicians, including Dee, Recorde, Digges, and Harriot, see Nicholas Popper, "The English Polydaedali: How Gabriel Harvey Read Late Tudor London," *Journal of the History of Ideas* 66 (2005): 351–81. For reference to "mathematiques," see Francis Shakelton, *A Blazying Starre or Burnyng Beacon* (London, 1580), EEBO, accessed April 11, 2011, A1r–v.

102. Johnson, *Astronomical Thought*, 120–25, discusses these early accounts of Copernican theory.

103. See J. L. Heilbron's introduction to Shumaker, *John Dee on Astronomy*, which helps the reader make some sense of this otherwise largely incomprehensible work.

104. On Harriot's career, his speculations on atomism, his correspondence with Kepler, and his knowledge of geometry and astronomy, see John W. Shirley, ed., *Thomas Harriot* (Oxford: Clarendon Press, 1974), esp. Jean Jacquot, "Harriot, Hill, Warner, and the New Philosophy," 107–28. For an edition of his mathematical writings, see Jacque-

line A. Stedall, *The Great Invention of Algebra: Thomas Harriot's Treatise on Equations* (Oxford: Oxford University Press, 2003). For a balanced account of the sources of Harriot's ideas, see Stephen Clucas, "Harriot and the Field of Knowledge in the Renaissance," in Fox, *Thomas Harriot.*

105. On the potential influence of Giordano Bruno on Harriot and his circle, see Jacquot, "Harriot," 108–10. Clucas, however, argues that Bruno did not significantly influence Harriot.

106. Johnson, *Astronomical Thought*, 124.

107. Ibid., 117.

108. Ibid., 161.

109. Ibid., 167.

110. Digges is loosely translating Copernicus here, who is also alluding to various "secrets" traditions: "For in this most beautiful temple, who would place this lamp in another or better position than that from which it can light up the whole thing at the same time? For the sun is not inappropriately called by some people the lantern of the universe, its mind by others, and its rules by still others. [Hermes] the Thrice Greatest labels it a visible god, and Sophocles' Electra the all-seeing." From Nicholas Copernicus, *On The Revolutions*, ed. and trans. Edward Rosen (Baltimore: Johns Hopkins University Press, 1978), 22.

111. Stephen Johnston, "Like Father, Like Son? John Dee, Thomas Digges, and the Identity of the Mathematician," in Clucas, *John Dee*, who argues that Digges "received no systematic grounding in academic philosophy," 78.

112. See Johnson, *Astronomical Thought*, 154–55.

113. Ibid., 78.

114. James M. Lattis, *Between Copernicus and Galileo: Christoph Clavius and the Collapse of Ptolemaic Astronomy* (Chicago: University of Chicago Press, 1974), 147.

115. Johnson, *Astronomical Thought*, 154.

116. Shumaker, *John Dee on Astronomy*, 161–63. "Nullum, ne uno quidem die (naturaliter) praeservaretur Individuum."

117. Johnson, *Astronomical Thought*, 154–55. On the method for using parallax to determine the distance of heavenly bodies, see Robert Goulding, "Wings (or Stairs) to the Heavens: The Parallactic Treatises of John Dee and Thomas Digges," in Clucas, *John Dee*, 41–45.

118. Harkness, *John Dee's Conversations*, 133–38.

119. Ibid., 68–69.

120. Ibid., 116.

121. For Harvey's life, see Virginia F. Stern, *Gabriel Harvey: His Life, Marginalia, and Library* (Oxford: Clarendon, 1979), 54–69, on the letters exchanged with Spenser, and 102–14, on the controversy with Nashe.

122. Stern, *Gabriel Harvey*, 70–71.

123. Indeed, William Fulke's *A Goodly Gallerye . . . of Meteors* published in 1563 provides naturalistic explanations for all such visions in the air. He explains visions of armies fighting in the sky as potentially caused by scenes happening on Earth reflected in a particular configuration of clouds in the sky.

124. Johnson, *Astronomical Thought*, 191. For an account of the whole controversy, see Stern, *Gabriel Harvey*, 15–32.

125. Francis Johnson takes Harvey seriously on this and points out a supposed error that Spenser makes in his description of the configuration of stars in the night sky (*Astronomical Thought,* 193–95).

126. Gerard Passannante has argued that Harvey's "pleasant and pithy familiar discourse, of the earthquake in Aprill last" was intended in part as "a kind of clever polemical response" to an earlier treatise published by William Golding that interpreted the earthquake as a divine omen. See Gerard Passannante, "The Art of Reading Earthquakes: On Harvey's Wit, Ramus's Method, and the Renaissance of Lucretius," *Renaissance Quarterly* 61 (2008): 800. I don't agree with his contention that the *De Rerum Natura* inspired Harvey's naturalistic approach to the earthquake. Passannante argues that Harvey cites a naturalistic passage from Ovid as a witty allusion to Golding's moralizing translation of the *Metamorphoses,* a citation that would otherwise be out of place among the philosophical sources that Harvey cites.

127. See Passannante, "Art," 794–96.

128. Popper, "English Polydaedali," 354.

129. Andrew Prewitt, "Gabriel Harvey and the Practice of Method," *Studies in English Literature, 1500–1900* 39 (1999): 26.

130. Alexander Grosart, ed., *The Works of Gabriel Harvey,* vol. 1 (1884; reprint, New York: AMS Press, 1966), 51. See Prewitt, "Gabriel Harvey," 26–27.

131. Grosart, *Works of Gabriel Harvey,* 43.

132. Prewitt, "Gabriel Harvey," 27.

133. See Clark, "Scientific Status," 211, for the Protestant idea that "since miracles had ceased and all created things were left with only their natural capacities, all causation must be natural." Examples of Protestant arguments to this effect include Reginald Scot's *Discoverie of Witchcraft* (1584), Samuel Harsnett's *Declaration of Egregious Popish Impostures* (1603), and William Fulke's *A Goodly Gallerye* (1563).

134. Johnson, *Astronomical Thought,* 191.

135. Although Stern notes that Gabriel Harvey accepted a fellowship in medicine at Pembroke College in 1584, she says that he educated himself in "modern scientific subjects: medicine, chemistry, pharmacology, and navigation" (*Gabriel Harvey,* 74).

136. Popper, "English Polydaedali," 352.

137. Ibid., 351–81, 381.

138. Ibid., 364.

139. A comment on Sig. a3v of Sacrobosco's *Textus de Sphaera,* as cited by Stern, *Gabriel Harvey,* 166.

140. Stern, *Gabriel Harvey,* 167, calls this passage "pre-Baconian" and gives a translation, which I have altered and completed.

141. Ibid., 166–67, Stern's translation: "Judicia mutanda: quia stellae fixae mutaverunt loca. Veteres Tabulas Ptolemaei, et aliorum, non congruere hodie cum Phaenomenis: patet maxime, ex Revolutionibus Copernici; et Tabulis Prutenicis Rheinoldi: unde antiquarum tabularum errors reformandi."

142. Smith, *Gabriel Harvey's Marginalia,* 127.

CHAPTER 3: Losing Touch with Nature

1. See William Eamon, *Science and the Secrets of Nature: Books of Secrets in Medieval and Early Modern Culture* (Princeton: Princeton University Press, 1994), 234, on

the "concept of science as a *venatio,* or a hunt," which seems related to these metaphors of travel. See also Piers Brown, "'That full-sail voyage': Travel Narratives and Astronomical Discovery in Kepler and Galileo," in James Dougal Fleming, *The Invention of Discovery, 1500–1700* (Burlington, VT: Ashgate, 2011), 15–28, on the relationship between Eamon's *venatio* and the metaphor of travel in Kepler and Galileo.

2. On Rastell's sources, see Elizabeth M. Nugent, "Sources of John Rastell's *The Nature of the Four Elements,*" *PMLA* 57 (1941): 74–88; and Johnstone Parr, "More Sources of Rastell's Interlude of the Four Elements," *PMLA* 60 (1945): 48–58. For the play itself, see Richard Axton, ed., "A new interlude and a mery of the nature of the four elementis," in *Three Rastell Plays* (Cambridge: D. S. Brewer, 1979), 31, ll. 29, 32, 34.

3. See Paul Slack, "Uses of Vernacular Medical Literature," in Charles Webster, ed., *Health, Medicine, and Mortality in the Sixteenth Century* (Cambridge: Cambridge University Press, 1979), 243–46.

4. Thomas Elyot, *The Castel of Health* (London, 1541), A4r, Early English Books Online, accessed August 23, 2010.

5. Elyot, *Castel,* 2v. Craig Martin, *Renaissance Meteorology, Pomponazzi to Descartes* (Baltimore: Johns Hopkins University Press, 2011), 6–10, on the fact that the Aristotelian elements "are not found in a pure form in nature but are always together " (6).

6. There were certainly some treatises published in this period that do not demonstrate the same kind of interest in epistemological questions. Some are reprints of earlier works. (Caxton's *Mirrour of the Worlde,* originally published in 1481, and itself a translation of an earlier French work, *L'image du monde,* was reprinted in England in 1529.) There are medical texts that simply provide recipes for herbal remedies or, like Andrew Boorde's *Breviary of Health* (1547), list diseases, causes, and treatments without much interest in problems of diagnosis.

7. James Bono, *The Word of God and the Languages of Man* (Madison: University of Wisconsin Press, 1995), and Peter Harrison, "'The Book of Nature' and Early Modern Science," in Klaas Van Berkel and Arjo Vanderjagt, eds., *The Book of Nature in Early Modern and Modern History* (Leuven: Peeters, 2006), 1–26.

8. Bono, *The Word,* 43, 130–33.

9. Martin, *Renaissance Meteorology,* 32, 42.

10. Ibid., 25.

11. Ibid., 26.

12. Francis R. Johnson and Sanford V. Larkey, "Robert Recorde's Mathematical Teaching and the Anti-Aristotelian Movement," *Huntington Library Bulletin* 7 (1935): 60.

13. Ibid., 77–87. Johnson and Larkey compare him to Peter Ramus and Francis Bacon.

14. Robert Recorde, *The Ground of Artes, Teaching the perfect Worke and practice of Arithmeticke, both in whole numbers and Fractions, after a more easie and exact sort, than hitherto hath beene set forth* (London, 1592), EEBO, accessed August 25, 2010. I have not used the first edition because the earliest editions available on EEBO are not easily legible.

15. Joy Easton, "The Early Editions of Robert Recorde's Ground of Artes," *Isis* 58 (1967): 515.

16. Richard W. Hadden, *On the Shoulders of Merchants: Exchange and the Mathematical Conception of Nature in Early Modern Europe* (Albany: State University of New York Press, 1994), 105.

17. On Recorde and place value, see Paula Blank, *Shakespeare and the Mismeasure of Renaissance Man* (Ithaca: Cornell University Press, 2006), 121.

18. Alfred W. Crosby, *The Measure of Reality: Quantification and Western Society, 1250–1600* (Cambridge: Cambridge University Press, 1997), 43.

19. Recorde, *Ground of Artes*, 33, 43.

20. See Linda Woodbridge, ed., *Money and the Age of Shakespeare: Essays in New Economic Criticism* (New York: Palgrave, 2003), introduction, 8–9, on Recorde's place in a plethora of books on mathematics and bookkeeping published in sixteenth-century England.

21. Robert Recorde, *The Urinal of Physick* (London, 1547), EEBO, accessed August 23, 2010.

22. See Dedre Gentner and Michael Jeziorski, "The Shift from Metaphors to Analogy in Western Science," in Andrew Ortony, ed., *Metaphor and Thought*, 2nd ed. (Cambridge: Cambridge University Press, 1993), 447–80, where they contrast an earlier alchemical use of analogy that "embraced similarity in all its forms in reasoning about the natural world" (471) with a more targeted mapping used by modern scientists beginning with Robert Boyle. Similarly, Lorraine Daston argues that Galileo's "vision of a reformed natural philosophy" led him to "all but exclude explanatory analogies from his scientific writings" (302), see Lorraine J. Daston, "Galilean Analogies: Imagination at the Bounds of Sense," *Isis* (1984): 302–10.

23. Michel Foucault, *The Order of Things: An Archaeology of the Human Sciences* (New York: Vintage Books, 1973), 21–22.

24. Robert Recorde, *The Castle of Knowledge* (London, 1556), EEBO, accessed June 18, 2008.

25. See Francis R. Johnson, *Astronomical Thought in Renaissance England: A Study of the English Scientific Writings from 1500 to 1645* (Baltimore: Johns Hopkins University Press, 1937; reprint, New York, Octagon Books, 1968), 122–24, on Digges's *Prognostication*. Leonard Digges's connection with Dee is mentioned by Thomas Digges in the preface to his *Alae seu scalae mathematicae*, A2r. He explains that his father died when Thomas was still a child, having asked his friend John Dee to act as a second father to Thomas. On this connection, see Francis R. Johnson, "The Influence of Thomas Digges on the Progress of Modern Astronomy in Sixteenth-Century England," *Isis* 1 (1936): 398–99.

26. Johnson, *Astronomical Thought*, 123.

27. Ibid., 123–24.

28. Leonard Digges, *A prognostication everlasting of right good effect . . .* (London, 1592), EEBO, accessed April 11, 2011. I'm citing the later edition, which includes Thomas Digges's addendum.

29. Johnson, *Astronomical Thought*, 117, notes that "all sixteenth- and seventeenth-century treatises on astronomical instruments such as astrolabes, cross-staffs, or celestial globes, assumed a geocentric system because they could not do otherwise." Johnson argues that Recorde was an advocate for Copernicanism although his book describes a Ptolemaic universe. I believe Recorde was actually uncertain about the extent to which Copernicus offered a plausible new model of the physical universe.

30. See E. A. Burtt, *The Metaphysical Foundations of Modern Science* (1931; reprint, Mineola, NY: Dover, 2003), 47–51, on the importance of perspective.

31. Nicholas Copernicus, *On the Revolutions*, ed. Edward Rosen (Baltimore: Johns Hopkins University Press, 1978). In book 5, he explains that "there are two reasons why a planet's uniform motion appears nonuniform: the earth's motion and the planet's own motion" (240).

32. Johannes Kepler, *Optics: Paralipomena to Witelo and Optical Part of Astronomy*, trans. William H. Donahue (Santa Fe: Green Lion Press, 2000).

33. Dee's *Monas Hieroglyphica* (1564) pursues a different, less naturalistic approach to understanding the universe, applying ideas from various secrets traditions, including Cabala. See Deborah Harkness, *John Dee's Conversations with Angels: Cabala, Alchemy, and the End of Nature* (Cambridge: Cambridge University Press, 1999), 77–90.

34. The extent of Dee's "occultism" or "Hermeticism" and the sources that influenced him have been controversial. I agree with Deborah Harkness and Nicholas Clulee in accepting a more naturalistic Dee, influenced as much by the writings of Roger Bacon as by the Hermetic writers mentioned by Frances Yates. For an account of the controversy, see Nicholas Clulee, "John Dee's Natural Philosophy Revisited," in Stephen Clucas, ed., *John Dee: Interdisciplinary Studies in Renaissance Thought* (Dordrecht: Springer, 2010), 23–37. See also Clulee, "Astrology, Magic, and Optics: Facets of John Dee's Early Natural Philosophy," *Renaissance Quarterly* 30 (1977): 647–50, on a "naturalistic Aristotelian phase in Dee's thought" (650).

35. Wayne Shumaker, ed. and trans., *John Dee on Astronomy: Propaedeumata Aphoristica (1558 and 1568)* (Berkeley and Los Angeles: University of California Press, 1978), aphorism 3, 122–23. I'm using Shumaker's English translation. The Latin reads, "Non solum ea Esse asserendum est, quae Actu in rerum natura sunt conspicua, notaque: sed et illa quoque quae quasi seminaliter, in naturae latebris, Extare, Sapientes docere possunt."

36. Ibid., 122–23: "Radios orbiculariter eiaculatur in singulas mundi partes, qui universum mundum suo modo replent."

37. John Dee, *Matheticall Praeface to the Elements of Euclid of Megara* (London, 1570), facsimile ed. (Whitefish, MT: Kessinger). In the "Preface," Dee is influenced by Platonic and Aristotelian ideas about mathematics and nature.

38. Leonard Digges and Thomas Digges, *A Geometrical Practise, named Pantometria* (London, 1571), EEBO, accessed April 13, 2011.

39. C. A. Ronan, "The Origins of the Reflecting Telescope," *Journal of the British Astronomical Association* 101 (1991): 335–42, argues that Leonard and Thomas Digges did build this device and that it constituted the first successful reflecting telescope. Ronan cites a letter written by William Bourne to Willam Cecil that mentions the achievements of Dee and the Diggeses in constructing devices that "shewe the thinge of marvelous largeness, in manner uncredable to bee believed of the common people" (Bourne letter, cited in Ronan, "Origins," 339). Other historians have questioned the claims made by Digges about the power of the device, arguing that the device he describes would not be capable of rendering coins legible from seven miles away.

40. William Fulke, A *Goodly Gallerye with a Most Pleasaunt Prospect, into the garden of natural contemplation, to behold the natural causes of all kynde of* Meteors, *as wel fiery, as watry and earthly* . . . (London, 1563), EEBO, accessed April 11, 2011.

41. Fulke originally published the *Antiprognosticon* in Latin. I am citing the 1560 translation by William Painter, *Antiprognosticon, that is to saye, an Invective against the*

vayne and unprofitable predictions of the Astrologians as Nostrodame, etc. (London, 1560), EEBO, accessed December 18, 2013. On this translation and Fulke in general, see Richard Bauckham, "Science and Religion in the Writings of Dr. William Fulke," *British Journal for the History of Science* 8 (1975): 18. Bauckham describes the educational board games developed by Fulke in the 1570s (19).

42. Stuart Clark, *Thinking with Demons: The Idea of Witchcraft in Early Modern Europe* (Oxford: Oxford University Press, 1997), 211, discusses the Protestant argument that miracles no longer occurred.

43. Bauckham, "Science," 21.

44. Fulke, *Antiprognosticon*, B8v.

45. Bauckham, "Science," 28, citing Trinity College, Dublin, MS 165, fol. 117r. The manuscript is attributed to "Wm F" and probably dates from around 1570.

46. Bauckham, "Science," 28. S. K. Heninger, *A Handbook of Renaissance Meteorology* (Durham, NC: Duke University Press, 1960), 21, notes that "Fulke's attitude may best be described as 'scientific' by Elizabethan standards: he sought rational explanations of the various irregularities in nature, and attempted to fit these apparent anomalies into a plan of natural order established by the Supreme Deity." Bauckham is concerned to establish Fulke as a precursor of the "scientific spirit," while Heninger emphasizes its religious context and the ways in which it falls short of modern scientific principles.

47. Johnson, *Astronomical Thought*, 154.

48. Ibid., 154.

49. Harkness, *John Dee's Conversations*, 134.

50. Ibid., 133–37. She notes that "Dee purchased nine works on the new star of 1572 and acquired the printed collection of letters exchanged by fellow Englishmen Gabriel Harvey and Edmund Spenser on the Easter earthquake of 1580" (135).

51. For the literary history of the trope of travel into the heavens, see Marjorie Hope Nicolson, *Voyages to the Moon* (New York: Macmillan, 1948), 1–21, who mentions Lucian, the *Somnium Scipionis*, and various medieval traditions, as well as marvelous voyages in Rabelais, Cervantes, and Ariosto.

52. Harkness, *John Dee's Conversations*, 60.

53. Ibid., 66–67.

54. The translation of the title is from David Riley and Stephen Pumfrey, "England's First Copernican: A New Text by Thomas Digges on the 'New Star' of 1572," Lancaster University reprints, available at eprints.lancs.ac.uk/969/, accessed March 11, 2011. The Latin text is from Thomas Digges, *Alae Seu Scalae Mathematicae, quibus visibilium remotissima Caelorum Theatra Conscendi, et Planetarum omnium itinera novis et inauditis Methodis explorari: tum huius portentosi Syderis in Mundi Boreali plaga insolito fulgore coruscantis, Distantia et Magnitudo immense, Situsque; protinus tremendous indagari, Deisque; stupendum ostentum, Terricolis expositum, cognosci liquidissime possit* (London, 1573), EEBO, accessed April 11, 2011.

55. Pumfrey and Riley, "England's First Copernican," 2. They argue that the image of "wings" is derived from Plato's *Phaedrus*, which calls arithmetic and geometry the "wings of the human mind," 4. Digges thus seems to be calling on both Platonic and biblical imagery.

56. My translation, A2r. "Longe remotiorem Cometarum locis esse perspexerit, cum in ipso Aethere purissimo ubi naturaliter nullae Mutationes, Innovationes, aut Alterationes

existere possint, quo nulla vi vapores Cometarum (Authore Aristotele omnibusque Physicis consentientibus) subiectum, evehi possunt."

57. "Adeo ut hoc ipsum rarissimum signum vere miraculum, Deique, nuntium, aut contemnant aut parvi existiment." My translation, A2r.

58. A2v, translation from Johnson, *Astronomical Thought,* 159. "An Terra immobilis in Mundi centro quiescat, et ingens illa Orbium erraticarum et fixarum moles rapidissimo cursu 24. Horarum spacio in gyrum rotetur, seu potius fixarum illa immensa sphaera vere fixa maneat, et apparens ille motus tantumodo ex Terre circulari super Polis suis rotatione contingat."

59. Francis Shakelton, *A Blazyng Starre or burnyng Beacon, seene the 10 of October laste (and yet continewyng) set on fire by Gods providence, to call all sinners to earnest and spedie repentance* (London, 1580), a3r, EEBO, accessed July 13, 2009.

60. Harkness, *John Dee's Conversations,* 134–35.

61. On Yates's ideas about Bruno and Hermeticism, and the questioning of those ideas by more recent historians of science, see Clulee, "John Dee's Natural Philosophy Revisited," 26–29.

62. See John Henry, "Thomas Harriot and Atomism: A Reappraisal," *History of Science* 20 (1982): 267–95. For the argument that Harriot's atomism was based on ancient sources, see Robert Kargon, *Atomism in England from to Newton* (Oxford: Clarendon, 1966).

63. See Stephen Clucas, "Harriot and the Field of Knowledge in the Renaissance," in Robert Fox, ed., *Thomas Harriot: An Elizabethan Man of Science* (Burlington, VT: Ashgate, 2000), 106.

64. On the place of this work in the history of Paracelsan writings in England, see Allen G. Debus, "The Paracelsan Compromise in Elizabethan England," *Ambix* 8 (1960): 77.

65. See Debus, "Parcelsan Compromise," on the attribution of authorship ("R. B." in the published text) to Richard Bostocke. Debus argues that Bostocke's book was not widely read. Most Paracelsan texts published in England in the sixteenth century consisted of recipes for chemical medicines and did not include explanation of Parcelsan theories. Debus argues that in sixteenth-century England many doctors simply added chemical medicines to a basically Galenic medical practice (96–97).

66. Richard Bostocke, *The difference betwene the auncient Phisicke . . .* (London, 1585), EEBO, accessed August 5, 2010.

67. See Amy Eisen Cislo, "Paracelsus's Conception of Seeds: Rethinking Paracelsus's Ideas of Body and Matter," *Ambix* 55 (2008): 274–82, who focuses on his ideas about the role of seeds in human conception.

68. See Clucas, "Harriot," 93–136. Clucas particularly cautions against "falling into the anachronistic binary opposition of 'occult' versus 'scientific'" (135). For the largely unsubstantiated arguments about Harriot's role in an occult "school of night," see M. C. Bradbrook, *The School of Night: A Study in the Literary Relationships of Sir Walter Raleigh* (Cambridge: Cambridge University Press, 1936); and Frances Yates, *A Study of Love's Labors Lost* (Cambridge: Cambridge University Press, 1936). For the argument that Harriot embraced a systematic atomic theory of matter, see Robert Kargon, "Thomas Hariot, the Northumberland Circle, and Early Atomism in England," *Journal of the History of Ideas* 27 (1966): 128–36. For a critique of Kargon's argument, see John Henry,

"Thomas Harriot and Atomism: A Reappraisal," *History of Science* 20 (1982): 267–96. On Harriot as an English Galileo, see Jean Jacquot, "Harriot, Hill, Warner and the New Philosophy," in J. W. Shirley, ed., *Thomas Harriot, Renaissance Scientist* (Oxford: Oxford University Press, 1974).

69. Hilary Gatti, "The Natural Philosophy of Thomas Harriot," in Fox, *Thomas Harriot*, 70. Gatti is mostly interested in charting the ways in which Harriot departed from Aristotelian tradition.

70. The English translation and explanation of the experiment are from Kargon, "Thomas Hariot," 26. Harriot wrote in Latin: "Ergo corpus densum diaphanum quod sensu videtur esse continuum per omnes partes, revera continuum non est. Sed habet partes corporeas quae radiis resistunt, et partes incorporeas radiis pervias. Ita ut refractio nihil aliud est quam interna reflexio, et pars radii intro recepta etsi videtur sensui esse recta, est tamen revera composita ex multis." From Johannes Kepler, *Gesammelte Werke*, vol. 15, *Briefe 1604–1607*, ed. Max Caspar, 25 vols. to date (Munich: C. H. Beck'sche, 1937), 15:367–68.

71. "Revera" is an intensification of "re" (in the sense "an actual thing, the thing itself, reality, truth, fact; opposed to appearance, mere talk, the mere name of a thing" but "strengthened by vera (sometimes as one word revera) in fact, really, in truth, indeed, in reality," see Charlton T. Lewis and Charles Short, *A Latin Dictionary* (Oxford: Clarendon Press, 1879; reprint, 1975), "Res," II A, 1575.

72. In Latin: "Jam duxi te ad fores domiciliorum naturae ubi latent eius arcana. Si non possis intrare propter illarum angustias, tum mathematice abstrahe et contrahe teipsum in atomum et intrabis facile. Et postquam egressus es dic mihi quae mirabilia vidisti" (Kepler, *Gesammelte Werke*, 15:368).

73. "Multa in tabella sum miratus, quae antea ignoravi" (Kepler, *Gesammelte Werke*, 16:31).

74. "Clavem mihi porrectam opinor ad inquirendas latentes corporum naturas" (Kepler, *Gesammelte Werke*, 16:31).

75. Kargon, "Thomas Hariot," 27.

76. "Si illae assumptiones et rationes tibi satisfaciunt, miror" and "Nullam Dyaphanitatem agnosco nisi ratione vacui" (Kepler, *Gesammelte Werke*, 16:172). First phrase translated in Kargon, "Thomas Hariot," 27. Second phrase, my translation.

77. Kepler, *Gesammelte Werke*, 16:173. "Nosti quod aurum inter elementaria nostra composita densissima habetur et opacum. Invenio tamen, cum corpus fit tenue, ut in vulgaribus foliis quibus pictores utuntur, quod sit dyaphanum. Si enim folium interponatur, inter oculum et candela lucentem, lux tota in propria forma perspicitur, sub viridi tamen colore."

78. See Edgar Zilsel, "The Origins of William Gilbert's Scientific Method," *Journal of the History of Ideas* 2 (1941): 1–32. Zilsel describes the *De Magnete* as the "first printed book, written by an academically trained scholar and dealing with a topic of natural science, which is based almost entirely on actual observation and experiment" (1).

79. Gad Freudenthal, "Theory of Matter and Cosmology in William Gilbert's *De Magnete*," *Isis* 74 (1983): 22. For an account of differing historical opinions about Gilbert, see Mary Hesse, "Gilbert and the Historians," *British Journal for the Philosophy of Science* 11 (1960): 1–10, 130–42.

80. Peter Dear, *Revolutionizing the Sciences: European Knowledge and its Ambitions,*

1500–1700 (Princeton: Princeton University Press, 2001), 52, 57. Dear sees Gilbert as like Thomas Digges in his reliance on "mixing of mathematical practice and natural philosophy" (55).

81. For Gilbert's relation to a poetic tradition of "world making" which has connections with craft knowledge, see Elizabeth Spiller, *Science, Reading, and Renaissance Literature: The Art of Making Knowledge, 1580–1670* (Cambridge: Cambridge University Press, 2004), 44–58.

82. Bruno Latour, *Pandora's Hope: Essays on the Reality of Science Studies* (Cambridge: Harvard University Press, 1999), 30.

83. Spiller, *Science*, 46, argues that Gilbert's "globe magnets thus represent not so much a thing as a way of creating knowledge through a certain kind of doing."

84. Freudenthal, "Theory," 23.

85. Ibid., 24, 23.

86. Ibid., 23.

87. William Gilbert, *On the Lodestone and Magnetic Bodies*, trans. P. Fleury Mottelay, in *Gilbert, Galileo, Harvey*, ed. Robert Maynard Hutchins and Mortimer Adler, vol. 28 of 54, Great Books of the Western World (Chicago: Encyclopedia Britannica, 1952), 1.

88. Peter Dear, *Revolutionizing the Sciences*, 55–57.

89. Spiller emphasizes the role of the terrella as a constructed world. Peter Dear, *Discipline and Experience: The Mathematical Way in the Scientific Revolution* (Chicago: University of Chicago Press, 1995), 158–61, argues that Gilbert's analogy between the terrella and the Earth is an example of a "mimetic" relationship between an artificial apparatus and nature.

90. Dear, *Discipline and Experience*, 160.

91. Ibid. Dear notes the role of this passage in Gilbert's argument.

92. See Keith Hutchison, "What Happened to Occult Qualities in the Scientific Revolution?," in *The Scientific Enterprise in Early Modern Europe: Readings from* Isis (Chicago: University of Chicago Press, 1997), 86–87, who argues that "occult qualities became fully and consciously accepted in natural philosophy, just as it became recognized that no qualities were ever directly perceived."

CHAPTER 4: Spenser and the New Science

1. A few scholars have linked the passages on precession and planetary motion in Book V and the *Mutabilitie Cantos* with the new science, although none has explored this in detail. See Judith Anderson, "Weighing Words With Spenser's Giant," in Peter Medine and Joseph Wittreich, eds., *Soundings of Things Done: Essays in Honor of S. K. Heninger, Jr.* (Newark: University of Delaware Press, 1997), 74, who cites "new stars, prophecies, and seemingly irregular planetary movements," as a context for Book V. See also Russell J. Meyer, "'Fixt in heavens light': Spenser, Astronomy, and the Date of the *Cantos of Mutabilitie*," *Spenser Studies* 4 (1983): 115–30, on Spenser's knowledge of astronomy.

2. Allison Chapman, "The Politics of Time in Edmund Spenser's English Calendar," *SEL* 42 (2002): 1–24.

3. See Richard Helgerson, "Two Versions of Gothic," in *Forms of Nationhood: The Elizabethan Writing of England* (Chicago: University of Chicago Press, 1992), 25–49,

on Spenser's use of classical and gothic models. See Teresa Krier, *Gazing on Secret Sights: Spenser, Classical Imitation, and the Decorums of Vision* (Ithaca: Cornell University Press, 1990), 8, for the argument that Spenser borrows from Virgil and Ovid in part because "the social and physiological dynamics of vision" are "central to their understanding of desire and human relation to the world."

4. On Spenser's engagement with ethical, political, and legal questions, see Elizabeth Fowler, "The Failure of Moral Philosophy in the Work of Edmund Spenser," *Representations* 51 (1995): 47–76. A number of scholars have discussed Spenser's relation to Neoplatonism, see esp., Jon Quitslund, *Spenser's Supreme Fiction: Platonic Natural Philosophy and* The Faerie Queene (Toronto: Toronto University Press, 2001). I tend to agree with Paul Suttie, "The Lost Cause of Platonism in *The Faerie Queene*," *Spenser Studies* 24 (2009): 413–30, who argues that Spenser's Platonic structures and episodes are almost always "called radically in doubt within the poem" (413). Many critics have read the poem primarily in relation to Spenser's relation to the English court, or to his implication in the colonization of Ireland, see esp. Louis Adrian Montrose, "The Elizabethan Subject and the Spenserian Text," in Patricia Parker and David Quint, eds., *Literary Theory/Renaissance Texts* (Baltimore: Johns Hopkins University Press, 1986), 303–40; Stephen Greenblatt, "To Fashion a Gentleman: Spenser and the Destruction of the Bower of Bliss," in *Renaissance Self-Fashioning from More to Shakespeare* (Chicago: University of Chicago Press, 1980), 157–92. On whether or not it is legitimate to read the poem through any lens other than the Irish one, see Andrew Hadfield and Willy Maley, "The Present State of Spenser Studies," in Jennifer Klein Morrison and Matthew Greenfield, eds., *Edmund Spenser: Essays on Allegory and Culture* (Aldershot, UK: Ashgate, 2000), 183–96. See Robert M. Schuler, "Theory and Criticism of the Scientific Poem in Elizabethan England," *English Literary Renaissance* 15 (1985): 36–37, for Harvey's remarks here in the context of a climate in late sixteenth-century England that was generally hostile to nonmimetic didactic poetry.

5. C. C. Moore Smith, ed., *Gabriel Harvey's Marginalia* (Stratford-Upon-Avon: Shakespeare Head Press, 1913), 159.

6. *Marginalia*, 162–63, my translation: "Pudet ipsum Spenserum, etsi Sphaerae, astrolabiique non plane ignarum; suae in astronomicis Canonibus, tabulis, instrumentisque imperitiae. Praesertim, ex quo vidit Blagravi nostri Margaritam Mathematicam. Qui ne Pontano quidem, aut Palingenio, aut Buchanano, aut etiam Bartasio cedit, exquisita utriusque Globi, astrolabii, baculique familiaris scientia."

7. Francis Johnson, *Astronomical Thought*, 193–95, has a favorable view of Harvey's level of scientific knowledge and accepts his assessment of Spenser's ignorance. Meyer, "Fixt in heavens light," 118–20, argues that Spenser knew more about astronomy than Harvey or Johnson suggest and that he refers to a specific eclipse of the moon in the *Cantos*.

8. See Deborah Harkness, *John Dee's Conversations with Angels* (Cambridge: Cambridge University Press, 1999), 68–69.

9. Edmund Spenser, *The Faerie Queene*, ed. A. C. Hamilton (London: Longman, 1977), VII.vi.13. All quotations from *The Faerie Queene* will be taken from this edition. Text references are to book, canto or proem, stanza, and line(s) of this edition.

10. Lauren Silberman, "*The Faerie Queene*, Book V, and the Politics of the Text," in *Spenser Studies* 19 (2004): 6.

11. Silberman, "*The Faerie Queene*, Book V," 6.

12. Spenser, *The Faerie Queene* (Longman, 1977), V. pr. 5. 1–5.

13. *The Faerie Queene*, V.pr.7. 4–8. The Variorum Spenser cites G. S. Hilliard as pointing out that " 'thirtie' is incorrect, but may be a misprint for 'thirteen,' which would be nearly correct," 158; see *The Works of Edmund Spenser, A Variorum Edition*, ed. Edwin Greenlaw, Charles Grosvenor Osgodd, Frederick Morgan Padelford et al., 11 vols. (Baltimore: Johns Hopkins University Press, 1932–49), vol. 5.

14. For Spenser's knowledge of contemporary debates about the calendar, see Chapman, "The Politics of Time," 7.

15. However, Paul Kocher, *Science and Religion in Elizabethan England* (San Marino, CA: The Huntington Library, 1953), 85, sees in this passage "a tendency among some Elizabethans to search among the sciences for whatever could be found to bolster an already existing mood of pessimism . . . In Donne the mood became an addiction and in Spenser it was frequent enough."

16. John Dee, *Mathematicall Praeface to the Elements of Geometrie of Euclid of Megara*, facsimile ed. (Whitefish, MT: Kessinger), B4r.

17. *The Faerie Queene*, V.ii.32.2. "Encroach" does not appear previously in *The Faerie Queene* and, apart from this episode, appears only once more, in Book VI.

18. Dee does disagree with Aristotle at one point, when he corrects "great Errors" in his theory of natural motion, refuting the idea that "Of any bodyes, the heavyer, to move downward faster then the lighter" (c1r). Dee was ahead of his time in recognizing what Galileo would later make more widely known.

19. Interestingly, Aristotle brings up the issue of commensurability in the *Nichomachean Ethics*, the work from which Spenser (at least nominally) takes the list of virtues around which his poem is organized. In his discussion of Justice and Equity, Aristotle notes that while money makes it possible to compare the worth of different kinds of goods in commercial exchanges (and therefore to determine equitable exchange values), this kind of commensurability does not extend to mathematics more generally: "in truth it is impossible that things differing so much should become commensurate." See Richard Hadden, *On the Shoulders of Merchants: Exchange and the Mathematical Conception of Nature in Early Modern Europe* (Albany: State University of New York Press, 1994), 79. Hadden argues that advances in early modern bookkeeping led to an abstraction of value and enabled thinkers like Galileo to begin including very different phenomena (distance and force) in the same equation. When Artegall questions the commensurability of right and wrong, he is maintaining an orthodox Aristotelian position.

20. Annabel Patterson, "The Egalitarian Giant," in *Reading Between the Lines* (Madison: University of Wisconsin Press, 1993), 90–91.

21. Paula Blank, *Shakespeare and the Mismeasure of Renaissance Man* (Ithaca: Cornell University Press, 2006), 158–59, traces the history of scales "as symbol of equity."

22. Bruno Kisch, *Scales and Weights, An Historical Outline* (New Haven: Yale University Press, 1956), 76–77, notes that although single arm balances were common in antiquity, justice is always symbolized by "a balance with two pans and an equal-arm beam" because "the equal arms of the balance represent more clearly the idea of equal treatment."

23. See Mary Phillips Berry, "On the Psychostasis in Christian Art," *Burlington Magazine for Connoisseurs* 22, no. 118 (Jan. 1913): 208–14.

24. Although some critics have supported Artegall's condemnation of the Giant, Elizabeth Fowler argues that Artegall does not speak for Spenser and that he uses the concept of equity as "the justification of his own power" ("Failure," 64).

25. Lorraine Daston, "Galilean Analogies: Imagination at the Bounds of Sense," *Isis* 75 (1984): 302–10; Dedre Gentner and Michael Jeziorsky, "The Shift from Metaphor to Analogy in Western Science," in A. Ortony, ed., *Metaphor and Thought,* 2nd ed. (Cambridge: Cambridge University Press, 1992), 447–80.

26. See Gad Freudenthal, "Theory of Matter and Cosmology in William Gilbert's *De Magnete,*" *Isis* 74 (1983): 37, who states that Gilbert's was "the first theory to unify the physics of the sublunary and superlunary realms," which is also what Mutabilitie is seeking to do.

27. Schuler, "Theory," 9.

28. Meyer, "Fixt in heavens light," argues that Spenser is referring to an eclipse that occurred on Monday, April 14, 1595, and therefore allows us to date the composition of the *Cantos* (126–27).

29. See Fowler, "Failure," 67–70, for the legal issues at stake here.

30. Spenser, *The Faerie Queene* (Longman, 1977), 726, note to VII, vii, 14, lines 7–9. A number of critics see the two episodes as linked, including Fowler, "Failure."

31. Hamilton, in Spenser, *The Faerie Queene* (Longman, 1977), notes Aristotle, *Meteorologica,* I.iii.339b, as a potential source.

32. Meyer, "Fixt in heavens light," 123.

33. Cited in Virginia Stern, *Gabriel Harvey: His Life, Marginalia, and Library* (Oxford: Oxford University Press, 1979), 166–67. The note is found in the margins of his copy of Firminius, *Reportorium de Aeris mutatione,* and reads in Latin: "*Judicia mutanda*: quia stellae fixae *mutaverunt* loca. Veteres Tabulas Ptolemaei, et aliorum, *non congruere hodie cum Phaenomenis*: patet maxime, ex *Revolutionibus Copernici*; et *Tabulis Prutenicis* Rheinoldi: unde antiquarum tabularum errors reformandi."

34. Critics are divided as to whether Nature's judgment is valid or not. Gordon Teskey, "Thinking Moments in *The Faerie Queene,*" *Spenser Studies* 22 (2007), argues that she bases her decision not on visible evidence but on a judgment about Mutabilitie's pushy attitude (122–23). Others, like Fowler, "Failure," 70, and Glen A. Steinberg, "Chaucer's Mutability in Spenser's *Mutabilitie Cantos,*" *SEL* 46 (2006): 37, see Nature's judgment as inadequate.

35. See Louise Gilbert Freeman, "Vision, Metamorphosis, and the Poetics of Allegory in *The Mutabilitie Cantos,*" *SEL* 45 (2005): 66, for the contradiction between "unperfite" and "perfection"; see p. 37, for the narrator's expression of "confusion, dissatisfaction, and doubt" in the two concluding stanzas. Fowler argues that the narrator here "appeals the important questions raised by the trial to the trump court of Judgment Day, and thus to divine law" ("Failure," 70).

36. Freeman sees Nature's veil as a figure for allegory and argues that "the *Mutabilitie Cantos* is the work in Spenser's oeuvre in which his anxieties about the accessibility of the divine (or even the ideal) through the instrument of poetic allegory came closest to the surface" (66). I would agree but see Spenser's anxiety as focused on the accessibility of truth.

37. Steinberg argues that Chaucer's Nature is quite different from Spenser's figure. He notes that "for Spenser, disorder in the cosmos seems to be an inherent and pervasive

characteristic of the cosmos itself, beyond the capacity or understanding of human beings, but for Chaucer, disorder seems to be primarily a product of human influence, arising out of human pettiness and passion" (29). I would argue that this difference is in part a product of the changing ideas about the natural world in the sixteenth century.

38. Hamilton, gloss in Spenser, *The Faerie Queene* (Longman, 1977), 733.

39. Syrithe Pugh, "Acrasia and Bondage: Guyon's Perversion of the Ovidian Erotic in Book II of *The Faerie Queene*," in J. B. Lethbridge, ed., *Edmund Spenser: New and Renewed Directions* (Madison, NJ: Farleigh Dickinson University Press, 2006), 153, argues that this passages presents "the epistemological dispute between empiricism and Platonism." I do not agree that Platonism per se significantly informs this passage. I believe that contemporary debates about Aristotelian versus "secrets" traditions more generally provide a more compelling context.

40. William Eamon, *Science and the Secrets of Nature: Books of Secrets in Medieval and Early Modern Culture* (Princeton: Princeton University Press, 1994), 272–73.

41. Kocher, *Science*, 86, says that Spenser's "reference to a possible world in the moon, a consequence of the disintegration of the Aristotelian cosmology then beginning, was one of the earliest in English literature." He approves that Spenser is here not claiming that the universe is in a state of decay.

42. Eamon, *Science and the Secrets of Nature*, 270.

43. See Michael Schoenfeldt, "The Construction of Inwardness in *The Faerie Queene*, Book 2," in Patrick Cheney and Lauren Silberman, eds., *Worldmaking Spenser* (Lexington: University Press of Kentucky, 2000), 234–43. See also Harry Berger, Jr., *The Allegorical Temper: Vision and Reality in Book II of Spenser's Faerie Queene* (New Haven: Yale University Press, 1957), 66–67, who argues that *krasis* or proper blending of humors is important in Book II. James W. Broaddus, "Renaissance Psychology and the Defense of Alma's Castle," *Spenser Studies* 19 (2004): 135–58, focuses on the role of the heart within an "Aristotelian-Galenic body-soul relationship," in which "temperament, temperance, temperature, and temper are closely related concepts of mixture and balance with both psychic and physiological application" (137).

44. Thomas Elyot, *The Castel of Health* (London, 1541), 2r, EEBO, accessed August 23, 2010.

45. *The Touchstone of Complexions. Generallye applicable, expedient and profitable for all such, as be desirous & carefull of their bodylye health. Contayning Most easie rules and ready tokens, whereby every one may perfectly try, and throughly know, as well the exacte state, habite, disposition, and constitution, of his owne Body outwardly: as also the inclinations, affections, motions, & desires of his mynd inwardly. First written in Latine, by Levine Lemnie, and now englished by Thomas Newton* (London: 1576), 5r, EEBO, accessed April 11, 2011.

46. R. B. (usually identified as Richard Bostocke), *The Difference betweene the Auncient Phisicke . . . and the latter Phisicke* (London, 1585), C6v.

47. "Comely" appears more times in Book II (fourteen) than in any other book of *The Faerie Queene*. It appears ten times in Book III and ten times in Book VI. I would argue that this indicates a particular connection between comeliness and temperance (and secondarily with chastity and courtesy).

48. *The Faerie Queene*, II.ii.7.1–2. Helen Cooney, "Guyon and his Palmer: Spenser's Emblem of Temperance," *Review of English Studies*, new ser., 51 (2000): 172, notes that

there has been "virtual unanimity" that the Palmer "represents an aspect of reason, or more specifically, right reason, or the virtue of prudence."

49. Archimago presents clues that he is a questionable character despite his sage and sober appearance: his bead telling and breast knocking would be suspicious to Protestants, and he is described as "simple in shew" (I.i.29.7). However, the resemblance between the descriptions of the Palmer and Archimago, one good and one bad, is striking.

50. Cooney argues that the Palmer is "*himself* a hermeneut or interpreter" and functions as "a meta-figure for allegorical interpretation" ("Guyon," 170).

51. Anne Ferry, *The Art of Naming* (Chicago: University of Chicago Press, 1988), 47.

52. Teskey, "Thinking Moments," 107, and Lauren Silberman, "*The Faerie Queene*, Book II, and the Limitations of Temperance," *Modern Language Studies* 17 (1987): 9.

53. Schoenfeldt, "Construction," 239.

54. Broaddus, "Renaissance Psychology," 146–47, cites Helkiah Crooke's later echoing of this passage in his *Microcosmographia* (1615).

55. According to Hamilton's gloss in Spenser, *The Faerie Queene* (Longman, 1977), 256.

56. Robert L. Reid, "Alma's Castle and the Symbolization of Reason in *The Faerie Queene*," *Journal of English and Germanic Philology* 80 (1981): 523; and Berger, *Allegorical Temper*, 82. Reid and Berger disagree about the sources for Spenser's account of these faculties.

57. Lynette C. Black, "Prudence in Book II of *The Faerie Queene*," *Spenser Studies* 13 (1999): 71.

58. Ibid., 71–75, describes both traditions. For an account of the iconography of prudence, see Erwin Panofsky, *Meaning in the Visual Arts* (Garden City, NY: Doubleday, 1955; reprint, Chicago: University of Chicago Press, 1982), 146–68.

59. Accounts of faculty psychology, with minor variations in detail, can be found in many sources. For a modern summary, see William Rossky, "Imagination in the English Renaissance: Psychology and Poetic," *Studies in the Renaissance* 5 (1958): 50–51.

60. Berger, *Allegorical Temper*, 78, 81.

61. Cooney, "Guyon," 172. And see Black, "Prudence," 65.

62. Black explains this with reference to the iconography of prudence: "as the second face of prudence which looks to the present, that most elusive of all moments, the brevity of the visit is accounted for by the brief poise of present ripeness, another attribute of prudence" ("Prudence," 72). I would argue that the brevity of the visit also stems from Spenser's uncertainty about the possibility of judging in the present moment.

63. Berger, *Allegorical Temper*, 83.

64. Stern, *Gabriel Harvey*, 167, calls this passage "pre-Baconian" and gives a translation, which I have altered and completed.

65. Teskey, "Thinking Moments," 121.

CHAPTER 5: Shakespeare and New Forms of Nothing

1. See my essay, "Optics," in Henry S. Turner, ed., *Early Modern Theatricality* (Oxford: Oxford University Press, 2013), 250–69.

2. Gail Kern Paster, *Drama and the Disciplines of Shame in Early Modern England* (Ithaca: Cornell University Press, 1993), and *Humoring the Body: Emotions and the Shakespearean Stage* (Chicago: University of Chicago Press, 2004).

3. S. K. Heninger, *A Handbook of Renaissance Meteorology* (Durham, NC: Duke University Press, 1960), 204, argues that "Shakespeare was quite knowledgeable in Aristotelian meteorology and his works abound in the technical terminology associated with it." See also Rebecca Totaro, "The Meteorophysiology of the Curse in Shakespeare's First Tetralogy," *English Language Notes* 51, no. 1 (2013): 191–210.

4. Howard Marchitello, *The Machine in the Text: Science and Literature in the Age of Shakespeare and Galileo* (Oxford: Oxford University Press, 2012), traces scholarly controversy over the significance for *Hamlet* of the appearance of Rosencrantz and Guildenstern among the names of ancestors listed in Tycho Brahe's *Astronomiae Instauratae Mechanica* (1598). Oxfordian Peter Usher implausibly claims that the play is an allegory about Copernican astronomy ("Advances in the Hamlet Cosmic Allegory," *Oxfordian* 5 (2001): 25–49), while others argue that the names were common and do not suggest a connection between Shakespeare and Brahe. I do not believe there are any unambiguous and direct references to the heliocentric model of the universe in Shakespeare's plays.

5. See Linda Woodbridge, introduction to *Money and the Age of Shakespeare: Essays in New Economic Criticism*, ed. Linda Woodbridge (New York: Palgrave, 2003), 1–18; Natasha Korda, "Dame Usury: Gender, Credit, and Accounting in the Sonnets and *The Merchant of Venice*," *Shakespeare Quarterly* 60 (2009): 129–53, 246; Patricia Parker, "Cassio, Cash, and the 'Infidel o': Arithmetic, Double-Entry Bookkeeping, and *Othello's* Unfaithful Accounts," in Jyotsna Singh, ed., *A Companion to the Global Renaissance: English Literature and Culture in the Era of Expansion* (Oxford: Blackwell, 2009), 223–41. Paula Blank, "Shakespeare's Social Arithmetics: Checking the Math of King Lear," in *Shakespeare and the Mismeasure of Renaissance Man* (Ithaca: Cornell University Press, 2006), 118–52, establishes Shakespeare's interest in practical and ontological implications of mathematical equivalence.

6. See "Specifying Unknown Things: The Algebra of *The Merchant of Venice*," in Bronwen Wilson and Paul Yachnin, eds., *Making Publics in Early Modern Europe: People, Things, Forms, of Knowledge* (New York: Routledge, 2010), 212–31. For the relationship between zero and death, see "Death by Numbers: Counting and Accounting in *The Winter's Tale*," in Diana Henderson, ed., *Alternative Shakespeares 3* (New York: Routledge, 2008), 158–80.

7. For Shakespeare's references to "atomies," see Jonathan Gil Harris, "Atomic Shakespeare," *Shakespeare Studies* 30 (2002): 47–51.

8. Katherine Duncan-Jones, ed., *Sir Philip Sidney* (Oxford: Oxford University Press, 1989), 246.

9. Lars Engle, "Afloat in Thick Deeps: Shakespeare's Sonnets on Certainty," *PMLA* 104 (1989): 832–43; Korda, "Dame Usury"; Thomas Greene, "Pitiful Thrivers: Failed Husbandry in the Sonnets," *Shakespeare and the Question of Theory*, ed. Patricia Parker and Geoffrey Hartman (New York: Methuen, 1985), 230–44; Peter C. Herman, "What's the Use? Or, the Problematic Economy in Shakespeare's Procreation Sonnets," in Peter Schiffer, ed., *Shakespeare's Sonnets: Critical Essays* (New York: Routledge, 2000), 263–83. Heather Dubrow, "'Incertainties now crown themselves assur'd': The Politics of Plotting in Shakespeare's Sonnets," *Shakespeare Quarterly* 47 (1996): 294, suggests that there are more "indeterminacies about plot and direction of address" in the sonnets than critics have acknowledged.

10. Christopher Martin, *Policy in Love: Lyric, and Public in Ovid, Petrarch, and Shakespeare* (Pittsburgh: Duquesne University Press, 1994), 147–48.

11. Stephen Booth, ed., *Shakespeare's Sonnets* (New Haven: Yale University Press, 1977). All quotations from Shakespeare's sonnets are from this edition.

12. Margaret J. Osler, "Galileo, Motion, and Essences," in Peter Dear, ed., *The Scientific Enterprise in Early Modern Europe: Readings from* Isis (Chicago: University of Chicago Press, 1997), 107.

13. See Reviel Netz, "Counter Culture: Towards a History of Greek Numeracy," *History of Science* 40 (2002): 321–52.

14. Richard W. Hadden, *On the Shoulders of Merchants: Exchange and the Mathematical Conception of Nature in Early Modern Europe* (Albany, NY: State University of New York Press, 1994), 67.

15. Ibid., 67.

16. Ibid., 64–65.

17. Jacob Klein, *Greek Mathematical Thought and the Origin of Algebra,* trans. Eva Brann (Cambridge: MIT Press, 1968; reprint, New York: Dover, 1992), 46–60. Diophantus did develop a form of algebra in antiquity, but, as Shankar Raman has argued in "Specifying Unknown Things," in Diophantus the unknown is always a single "determinate number" (217), while the sixteenth-century mathematician Francois Viète developed an algebra in which the unknown was "entirely indeterminate" (219). Raman ties to the algebra of Viète a process of abstraction similar to the one that I'm associating with the introduction of Arabic numerals.

18. See Dorothy V. Schrader, "The Arithmetic of the Medieval Universities," in Frank Swetz, ed., *The European Mathematical Awakening: A Journey Through the History of Mathematics from 1000 to 1800* (Mineola, NY: Dover, 2011), 23, for the idea that the mathematics taught in medieval universities was comparable to what is taught today in the eighth grade.

19. See Parker, "Cassio," on Arabic contributions to early modern mathematics and its role in commerce.

20. Hadden, *On the Shoulders,* 104–9, on Robert Recorde and the "golden rule."

21. Robert Recorde, *The Ground of Artes* (London, 1596), 43, EEBO, accessed August 25, 2010. I'm using a later edition because the earlier editions available on EEBO are not easily legible.

22. Robert Recorde, *The Whetstone of Witte* (London, 1557), bivr, EEBO, accessed April 11, 2011.

23. Mark Netzloff, "The Lead Casket: Capital, Mercantilism, and *The Merchant of Venice,*" in Woodbridge, *Money,* 162.

24. Netzloff, "The Lead Casket," 162.

25. Woodbridge, introduction to *Money,* 4. Korda, "Dame Usury," and Parker "Cassio," have also emphasized the new importance of mathematical knowledge in the period.

26. Greene, "Pitiful Thrivers," 231, 243.

27. Joel Fineman, *Shakespeare's Perjured Eye* (Berkeley and Los Angeles: University of California Press, 1986), 208, notes that "the procreation argument" is "a theme which is itself unprecedented in the genre of the sonnet sequence (although familiar enough, of course, in other literary contexts)."

28. On the various theories about procreation in the period, see Thomas Laquer, *Making Sex: Body and Gender from the Greeks to Freud* (Cambridge: Harvard University Press, 1990), 43–51; for a critique of Laquer's "one sex" argument, see Katherine Park and Robert Nye, "Destiny is Anatomy," *New Republic*, February 18, 1991, 53–57. For the idea that women also contributed "seed" to conception, see Elaine Hobby, ed., *The Birth of Mankind: Otherwise Named the Woman's Book* (Aldershot, UK: Ashgate, 2009), 37–43.

29. Engle, "Afloat," 832.

30. Fineman sees a Platonic source for the "unitary arithmetic of idealism" that allows us to accept that "the young man, multiplying himself after his own kind, will father the 'many' that will prove him 'one'" (251–52).

31. Herman, "What's the Use?" 270.

32. Ibid., 278.

33. Booth, *Shakespeare's Sonnets*, note on Sonnet 6, line 8, on p. 142.

34. Raman, "Death by Numbers," 158–59. In this case the power of zero and multiplication is associated with female sexuality and is therefore imagined as threatening.

35. Korda, "Dame Usury," 134–36; Parker, "Cassio," 227.

36. Sonnet 136, ll. 8–11.

37. Sonnet 135, l. 14.

38. Booth, *Shakespeare's Sonnets*, 462, traces some of the many puns in these lines, for instance on "*nothing* (1) worthless, a zero; (2) nonexistent, a 'non-thing.' There is also a play on 'no thing,' i.e. 'not a penis.'" He also notes the "hectic display of bawdy puns on words like *will* and *thing*."

39. Helen Vendler, *The Art of Shakespeare's Sonnets* (Cambridge: Harvard University Press, 1997), 72.

40. For a different approach to Shakespeare's mathematical interests in *Lear*, see Paula Blank, "Shakespeare's Social Arithmetics," 118–52. She traces Shakespeare's interest in Arabic numerals and the cipher (122) but argues that he is concerned in the play with suggesting "that Renaissance mathematics does not provide all the answers and that the human costs of miscalculation may prove immeasurable, more than numbers can tell" (120).

41. See Richard Halpern, "*Historica Passio*: *King Lear*'s Fall into Feudalism," in *The Poetics of Primitive Accumulation: English Renaissance Culture and the Genealogy of Capital* (Ithaca: Cornell University Press, 1991), 219. Halpern cites, as examples of such readings of the play, John Danby, *Shakespeare and the Doctrine of Nature* (London: Faber and Faber, 1952), and John Turner, "The Tragic Romances of Feudalism," in Graham Holderness, Nick Potter, and John Turner, eds., *Shakespeare: The Play of History* (Iowa City: University of Iowa Press, 1987), among others. For the atheism of the play, see Steven Greenblatt, "Shakespeare and the Exorcists," in *Shakespearean Negotiations* (Berkeley and Los Angeles: University of California Press, 1988), 119: "*King Lear* is haunted by a sense of rituals and beliefs that are no longer efficacious, that have been *emptied out*. The characters appeal again and again to the pagan gods, but the gods remain utterly silent."

42. Danby, *Shakespeare*, does place the play in relation to a shift from "the benignant nature" of Bacon and Hooker to "the malignant nature of Hobbes" (20, 31), but he doesn't go beyond a general account of the familiar ordered Elizabethan world picture to consider specific theories of matter.

43. Henry S. Turner, "*King Lear* Without: The Heath," *Renaissance Drama* 28 (1997): 177. Turner goes on to note that the play may trace "a moment of transition in spatial thinking . . . between a neo-Aristotelian scholastic philosophy that could conceive *only* of container or 'place' and the emergence of a distinct notion of 'space' understood as a homogeneous, extended medium that precedes and receives all bodies and their movements" (177–78).

44. See Thomas Shirley, *Thomas Harriot: A Biography* (Oxford: Clarendon, 1983), 327–79, on the years 1605–1607 and the Northumberland circle. See also John Henry, "Thomas Harriot and Atomism: A Reappraisal," *History of Science* 20 (1982): 267–96, for a careful account of the nature of Harriot's atomism. Although Harriot published virtually none of his scientific writings during his lifetime, they exist in manuscript. His atomistic theories can be inferred from a treatise written by his friend, Nathaniel Torporley, *A synopsis of the controversie of atoms* (British Library, London, Birch MS. 4458).

45. Henry, "Thomas Harriot," 267.

46. Ibid.; Kenelm Digby, *Two Treatises. In the one of which The Nature of Bodies in the other the Nature of Mans Soule, is looked unto* (London, 1641), 17.

47. Dedre Gentner and Michael Jeziorski, "The Shift from Metaphor to Analogy in Western Science," in Andrew Ortony, ed., *Metaphor and Thought* (Cambridge: Cambridge University Press, 1993), 456–71. I would separate the Aristotelian tradition (which gives little role in natural philosophy to what cannot be seen) from that of alchemical and other "secret" traditions that seek causal explanations for the unseen.

48. Ibid., 440–51.

49. *King Lear*, 4.6.52, 49.

50. See Shirley, *Thomas Harriot*, chaps. 7 and 8, for references to Harriot in those trials. See also Stephen Clucas, "Corpuscular Matter Theory in the Northumberland Circle," in Christoph Luthy, John E. Murdoch, and William R. Newman, eds., *Late Medieval and Early Modern Corpuscular Matter Theories* (Leiden: Brill, 2001), who argues that Warner, Hill, and Harriot "represent a vital transitional stage between Aristotelian concepts of matter and form in the sixteenth century and the later mechanical corpuscular theories of Hobbes, Boyle, and Newton in the later seventeenth century" (182).

51. John Aubrey, *Brief Lives*, ed. Richard Barber (Totowa, NJ: Barnes and Noble, 1975), 126–27.

52. *The Norton Shakespeare*, ed., Stephen Greenblatt (New York: W. W. Norton, 1997), 2322; *Riverside Shakespeare*, 1256.

53. In *Macbeth*, "though the treasure / Of nature's germains tumble all together, / Even till destruction sicken." In *Antony and Cleopatra*: "th' air, which, but for vacancy. / Had gone to gaze on Cleopatra too." All citations from Shakespeare's plays are from G. Blakemore Evans, ed., *The Riverside Shakespeare* (Boston: Houghton Mifflin, 1974). The orthodox Aristotelian idea that "vacancy" is impossible is here attributed to Enobarbus. It does not necessarily mean that Shakespeare accepted this doctrine, but it does suggest that he was aware of the issue.

54. E. J. Dijksterhuis, *The Mechanization of the World Picture* (Oxford: Clarendon Press, 1961), 24.

55. Digby, *Two Treatises*, 9.

56. Cited in Henry, "Thomas Harriot," 270, who also explains Harriot's analogy

between material particles and geometric points. Harriot's notes on Zeno are found in British Library, London, Add. MS.6782b, folio 363r.

57. Henry, "Thomas Harriot," 271.

58. See John E. Murdoch, "The Tradition of *Minima Naturalia*," in Luthy, Murdoch, and Newman, *Late Medieval*, 91–132.

59. As cited in Antonio Clericuzio, *Elements, Principles, and Corpuscles: A Study of Atomism and Chemistry in the Seventeenth Century* (Dordrecht: Kluwer, 2010), 76–77, from Nicholas Hill, *Philosophia Epicurea, Democritana, Theophrastica proposita simpliciter, non edocta* (Paris, 1601).

60. The English translation and explanation of the experiment are from Robert Kargon, *Atomism in England from Hariot to Newton* (Oxford: Clarendon, 1966), 26. Harriot wrote in Latin. From Johannes Kepler, *Gesammelte Werke*, 25 vols. to date, ed. Max Caspar and Walther Von Dyck (Munich: C. H. Beck'sche, 1937), 15:367–68.

61. "Si illae assumptions et rationis tibi satisfaciunt, moror" and "Nullam Dyaphanitatem agnosco nisi ratione vacui," Kepler, *Gesammelte Werke*, 16:172. First phrase translated in Kargon, *Atomism*, 27. Second phrase my translation.

62. See Clucas, "Corpuscular Matter," 182–95, on the influence of medieval optical theories of "radiative virtue" on the atomism of Harriot, Warner, and Hill.

63. Rosalie Colie, *Paradoxica Epidemica: The Renaissance Tradition of Paradox* (Princeton: Princeton University Press, 1966), 222.

64. Clericuzio, *Elements, Principles*, 76, cites Henry Percy, *Advice to his Son*, ed. G. B. Harrison (London, 1920), 70.

65. Clericuzio, *Elements, Principles*, 77.

66. Christoph Meinel, "Early Seventeenth-Century Atomism: Theory, Epistemology, and the Insufficiency of Experiment," in Dear, *The Scientific Enterprise*, 186.

67. Cicero, *De Senectute*, trans. William Falconer, Loeb Classical Library (Cambridge: Harvard University Press, 1950).

68. *King Lear*, 1.1.136–7. See, e.g., Halpern, "*Historica Passio*," 220–22.

69. See Danby, *Shakespeare*, for this view. It is critiqued by both John Turner, "Tragic Romances," 106–11, and Halpern, "*Historica Passio*," 240.

70. *King Lear*, 1.1.56. Henry Turner, "*King Lear* Without," 172, argues that the sisters here use "spatial rhetoric" and that "the association between the language of space, geometry, and betrayal is in keeping not only with other plays attributed to Shakespeare, but with a more pervasive mistrust of specialized figures and symbols in early modern Europe." I choose to emphasize the abstraction of their language rather than its spatial referents.

71. When Reagan says, "And find I am alone felicitate / In your dear Highness' love" (1.1.75–76), she does not say that he has all of her love but that only his love makes her happy.

72. This insight was pointed out to me by Steven Berkowitz in a tutorial at Harvard in 1978.

73. Shakespeare habitually contrasts "substance" and "shadow," e.g.: "Love like a shadow flies, when substance love pursues" (*Merry Wives*, 2.2.215); "so far this shadow / Doth limp behind the substance" (*The Merchant of Venice*, 3.2.128); "That are the substance / Of that great shadow I did represent" (2 *Henry VI*, 1.50.26); "for the very substance of the ambitious is merely the shadow of a dream" (*Hamlet*, 2.2.265).

74. See Gary Taylor, "The War in *King Lear*," *Shakespeare Survey* 33 (1980): 27, for the idea that this civil war is raised as the expected main focus of the play, an expectation that is then thwarted. See Marion Trousdale, "A Trip Through The Divided Kingdoms," *Shakespeare Quarterly* 37 (1986): 220–21, for a critique of Taylor's argument.

75. Henry Turner, "*King Lear* Without," 177.

76. See Clericuzio, *Elements, Principles*, 13, for various ancient and early modern theories of *semina* in relation to atomism. On this passage, see T. W. Baldwin, "Nature's Moulds," *Shakespeare Quarterly* 3 (1952): 237–41, who traces a series of passages that reveal Shakespeare's concept of the Earth as a womb, pregnant with seeds that will bring forth the future (unless destroyed, to revert to an original state of chaos). William Elton, "Timothy Bright and Shakespeare's Seeds of Nature," *Modern Language Notes* 65 (1950): 196–97, identifies a passage in Timothy Bright's *Treatise of Melancholie* that expresses a similar idea: that "this earth he, God, hath endued with a fecunditie of infinite seeds of all things: which he commaunded as a mother to bring forth."

77. Edward Taylor, "*King Lear* and Negation," *ELR* 20 (1990): 31, connects nothingness in *Lear* with "what lies between a maid's legs, as when Hamlet brutally jokes to Ophelia of 'country matters.'"

78. Turner, "*King Lear* Without," 167.

79. Bruce R. Bartlett, "Bearing the 'Waight': Double-Entendre in *Richard III, Othello*, and *King Lear*," *Shakespeare Newsletter* 240 (Spring 1999): 7–8.

80. Meinel, "Early Seventeenth-Century Atomism," 192.

81. Robert Heilman, *This Great Stage: Image and Structure in King Lear* (Baton Rouge: Louisiana State University Press, 1948; reprint, Seattle: University of Washington Press, 1963), 14–18.

82. Eve Sweetser, *From Etymology to Pragmatics* (Cambridge: Cambridge University Press, 1990), 39–40.

83. Ibid., 38.

84. Ibid., 37.

85. See Vivian Nutton, "The Seeds of Disease: An Explanation of Contagious Disease and Infection from the Greeks to the Renaissance," *Medical History* 27:1 (1983), esp. 20–34.

86. Meinel, "Early Seventeenth-Century Atomism," 186, citing Daniel Sennert's *Hypomnemata physica* (1636).

87. Halpern, "*Historica Passio*," 220.

88. Greenblatt, "Shakespeare and the Exorcists," 113, 119, 126, 127.

89. Halpern, "*Historica Passio*," 256. Halpern takes the phrase "Sum over histories" from physics, specifically from Richard Feynman's *QED: The Strange Theory of Light and Matter* (Princeton: Princeton University Press, 1985). Feynman interestingly begins to illustrate quantum mechanics using the problem of the partial reflection and refraction of light through an opaque surface, a problem that exercised seventeenth-century atomists and about which Harriot corresponded with Kepler.

CHAPTER 6: Matter and Power

1. E. M. W. Tillyard, *The Elizabethan World Picture* (New York: Vintage, 1959), 10; Marjorie Hope Nicolson, *The Breaking of the Circle: Studies in the Effect of the New Science Upon Seventeenth Century Poetry* (Evanston, IL: Northwestern University Press,

1950), 6. For a more recent account of the implication of botany in social hierarchy in the period, see Jean Feerick, "Botanical Shakespeares: The Racial Logic of Plant Life in *Titus Andronicus*," *South Central Review* 26 (2009): 82–102.

2. Caroline Spurgeon, *Shakespeare's Imagery and What It Tells Us* (Cambridge: Cambridge University Press, 1935), 13.

3. Marjorie Hope Nicholson, *Science and Imagination* (Ithaca: Cornell University Press, 1956; second printing, 1962), 41–42.

4. See above, introduction and chapters 2 and 3 above. Gabriel Harvey mentions Copernicus several times in his marginalia; see Virginia Stern, *Gabriel Harvey: His Life, Marginalia, and Library* (Oxford: Clarendon, 1979), 166–67. The evidence for Harriot's Copernicanism comes chiefly from a reference to it in a letter to Harriot from his friend Sir William Lowrer; see Matthias Schemmel, *The English Galileo: Thomas Harriot's Work on Motion as an Example of Preclassical Mechanics*, vol. 1 (Dordrecht: Springer, 2008), 20–21.

5. John Gillies, "Tamburlaine and Renaissance Geography," in Garrett Sullivan, Patrick Chency, and Andrew Hadfield, eds., *Early Modern English Drama: A Critical Companion* (New York: Oxford University Press, 2006).

6. Gillies, "Tamburlaine," 39–41, argues that "subversive force is wrung from the old cosmology by a relentless focus on its creaking joins," noting the "restless" spheres but not the new naturalistic meteorology or supralunar change. David Riggs, *The World of Christopher Marlowe* (New York: Henry Holt, 2004), 166–69, argues that "Copernican astronomy had raised the Ptolemaic system to new heights" by suggesting that man was capable of mathematically comprehending the universe. Riggs is mistaken, however, in arguing that Marlowe derived an Epicurean physics from his reading of Ovid's *Metamorphoses* (88–89).

7. See Francis Johnson, *Astronomical Thought in Renaissance England* (Baltimore: Johns Hopkins University Press, 1937; reprint, New York: Octagon Books, 1968), 94–96.

8. See Gillies, "Tamburlaine," and also Henry Turner, *The English Renaissance Stage: Geometry, Poetics, and the Practical Spatial Arts, 1580–1630* (Oxford: Oxford University Press, 2006), 6.

9. This connection was first made by Ethel Seaton, "Marlowe's Map," *Essays and Studies by Members of the English Association* 10 (1924): 13–35. For an exploration of the implications of these maps for Marlowe's implication in imperialism, see Gillies, "Tamburlaine."

10. Turner, *The English Renaissance Stage*, 6.

11. Gillies, "Tamburlaine," notes that Tamburlaine's ambition "unfolds in two dimensions—a geographic dimension and a cosmic dimension" (37)—but misses the role of the new meteorology and the supernova of 1572 in creating new possibilities for cosmic expansion.

12. David Fuller, ed., *Tamburlaine the Great*, in *The Complete Works of Christopher Marlowe*, vol. 5, ed. David Fuller and Edward J. Esche (Oxford: Clarendon Press, 1998). All quotations from *Tamburlaine* are taken from this edition.

13. S. K. Heninger, *A Handbook of Renaissance Meteorology* (Durham, NC: Duke University Press, 1960), 45–61.

14. Ibid., 128–34.

15. Arthur Golding, *A discourse upon the Earthquake* (London, 1580), EEBO, accessed June 27, 2012; Alexander Grosart, ed., *The Works of Gabriel Harvey*, vol. 1 (1884; reprint, New York: AMS Press, 1966); William Fulke, A *Goodly Gallerye with a Most Pleasaunt Prospect, into the garden of natural contemplation, to behold the natural causes of all kynde of* Meteors, *as wel fiery, as watry and earthly* . . . (London, 1563), EEBO, accessed April 11, 2011, ff. 22–23.

16. Gillies, "Tamburlaine," 40, notes that warring elements would suggest "illness" rather than "an aspiring mind."

17. On Tamburlaine's "mettle," or "spirited and courageous nature," see Mary Floyd-Wilson, "English Mettle," in Gail Kern Paster, Katherine Rowe, and Mary Floyd-Wilson, eds., *Reading the Early Modern Passions: Essays in the Cultural History of Emotion* (Philadelphia: University of Pennsylvania Press, 2004), 130.

18. Heninger, A *Handbook*, 176, reads this passage as Tamburlaine's "boast of continued sway over the sublunary universe," but "making the stars to melt" extends his destructive power above the moon.

19. Ibid., 177.

20. David Wootton, ed., *Doctor Faustus with the English Faust Book* (Indianapolis: Hackett, 2005), 3.3.Chorus.

21. For a comparison of Marlowe's and Shakespeare's relation to the atlas, see John Gillies, *Shakespeare and the Geography of Difference* (Cambridge: Cambridge University Press, 1994), 59–60.

22. Heninger, A *Handbook*, 204.

23. Ibid., 206. *Julius Caesar* is also full of meteorological portents, which were part of the inherited legends about Caesar's death.

24. "The Meteorophysiology of the Curse in Shakespeare's First Tetralogy," *English Language Notes* 51, no. 1 (2013): 191–210.

25. See, e.g., Norman Rabkin, *Shakespeare and the Common Understanding* (New York: Free Press, 1967), 186, 191, for an account of readings that contrast the values of Rome and Egypt. See also William D. Wolf, "'New Heaven, New Earth': The Escape from Mutability in *Antony and Cleopatra*," *Shakespeare Quarterly* 33 (1982), 328, for a characterization of the usual way the polarity has been read: Egypt as "regenerative, hot, emotional, the center of love and overripe sexuality"; Rome as "duty, public service, military valor, reason, and policy." See Jonathan Gil Harris, "'Narcissus in thy Face': Desire and the Difference It Fakes in *Antony and Cleopatra*," *Shakespeare Quarterly* 45(1994): 409, for an account of the controversial role of gender in this polarization of the play.

26. Gillies, *Shakespeare*, 118.

27. Ibid., 118.

28. G. Blakemore Evans, ed., *The Riverside Shakespeare*. Boston: Houghton Mifflin, 1974. All quotations from the play come from this edition.

29. See Donald C. Freeman, "'The Rack Dislimns': Schema and Metaphorical Pattern in *Antony and Cleopatra*," *Poetics Today* 20 (1999): 443–60, for a reading of the elemental language in the play in terms of the cognitive image schemas: CONTAINER, LINKS, and PATH. Like any dichotomy, my differentiation of Rome and Egypt breaks down. Toward the end of the play, for instance, Cleopatra adopts a consciously Roman

language when dealing with Caesar: "all the world, 'tis yours, and we / Your scutcheons and your signs of conquest" (5.2.134–35). Antony uses Egyptian and Roman language interchangeably.

30. William Cunningham, *The Cosmographicall Glasse, conteinyng the pleasant Principles of Cosmographie, Geographie, Hydrographie, or Navigation* (London, 1559), f. 6.

31. Shankar Raman, *Framing "India": The Colonial Imaginary in Early Modern Culture* (Stanford: Stanford University Press, 2002), 16, argues that a new sense of colonialist space was produced by developments in cartography and astronomy: "Out of this historical conjuncture comes a new figure: the abstract, geometrized spatial grid, itself allied with an understanding of Western reason as the universally valid form of rationality. Against the background of an ostensibly neutral and homogeneous space, 'India' and the 'East' take shape as *places*, as geographical regions 'produced' by concrete practices in accordance with specific needs."

32. See Leonard Tennenhouse, *Power on Display: The Politics of Shakespeare's Genres* (New York: Methuen, 1986), 146, for the idea that the play enacts a similar shift in the representation of political power, showing "that a whole way of figuring political power has been rendered obsolete." He calls the play "an elegy for the signs and symbols which legitimated Elizabethan power."

33. Gillies, *Shakespeare*, 4.

34. Raman, *Framing "India,"* 16.

35. For the link between colonialism and the new science, see Denise Albanese, *New Science, New World* (Durham: Duke University Press, 1996), esp. "Admiring Miranda, Enslaving Nature," 59–91.

36. For the division between the "two cultures" of science and literature, see C. P. Snow, *The Two Cultures*, reissue ed. (Cambridge: Cambridge University Press, 1993).

37. Gillies, *Shakespeare*, 117.

38. Stephen Toulmin and June Goodfield, *The Architecture of Matter* (Chicago: University of Chicago Press, 1962), 87. See Gail Kern Paster, "Melancholy Cats, Lugged Bears, and Early Modern Cosmology: Reading Shakespeare's Psychological Materialism Across the Species Barrier," Paster, Rowe, and Floyd-Wilson, eds., *Reading the Early Modern Passions*; and Floyd-Wilson, "English Mettle," 130–46, in the same volume, for the interaction of the human humoral system with its environment.

39. Bruno Latour, *Pandora's Hope: Essays on the Reality of Science Studies* (Cambridge: Harvard University Press, 1999), 170.

40. Elizabeth Spiller, "Shakespeare and the Making of Early Modern Science: Resituating Prospero's Art," *South Central Review* 26 (2009), 24–25.

41. Mary Floyd-Wilson, *English Ethnicity and Race in Early Modern Drama* (Cambridge: Cambridge University Press, 2003), 1–2. See her essay "English Epicures and Scottish Witches," *Shakespeare Quarterly* 57 (2006), 133, for the concept of "possibility," the idea that "people were embedded in extended circles of shaping forces that included one's diet, one's family, the state, natural environment, and the cosmological spheres."

42. Floyd-Wilson, *English Ethnicity*, 47.

43. Janet Adelman, *Suffocating Mothers: Fantasies of Maternal Origin in Shakespeare's Plays,* Hamlet *to* The Tempest (New York: Routledge, 1992), 177.

44. See Gail Kern Paster, *The Body Embarrassed: Drama and the Disciplines of Shame in Early Modern England* (Ithaca: Cornell University Press, 1993), 23–63.

45. As Henry Harris suggests, spontaneous generation was both believed and doubted from antiquity until it began to be disproved in the seventeenth century, first by Francesco Redi who demonstrated that rotten meat would not generate maggots unless exposed to flies; see Harris, *Things Come to Life: Spontaneous Generation Revisited* (Oxford: Oxford University Press, 2002). Aristotle wasn't clear about exactly how it worked in material terms, arguing at one point that the heat of the sun was the crucial factor but later arguing that some form of divine "pneuma" or "psyche" was necessary for life; see Toulmin and Goodfield, *Architecture*, 88–89. Spontaneous generation is thus at the heart of questions about material change, the nature of matter, and the interrelationship of matter and spirit.

46. Claude Nicolet, *Space, Geography, and Politics in the Early Roman Empire* (Ann Arbor: University of Michigan Press, 1991), 2.

47. Ibid., 20.

48. Ibid., 36.

49. Raman, *Framing "India,"* 90, 97.

50. See Ania Loomba, *Gender, Race, Renaissance Drama* (Manchester: Manchester University Press, 1989), 127, who argues that "whether the fight should take place on the Roman element, the land, or Cleopatra's medium, the water, is at once a matter of military strategy and a measure of Antony's emotional and political affiliations."

51. Freeman, "The Rack Dislimns," 457–58, notes the significance of this elemental dissolution and its accompanying failure of vision in terms of cognitive image schemas.

52. Jonathan Gil Harris, "Narcissus," 417.

53. See chapter 5 above, on the fact that early modern atomic theory raised the threatening possibility that void space or a vacuum could exist, contrary to Aristotelian teaching.

54. See, e.g., Paul Yachnin's analysis of the disjunction between Caesar's language of command and the loyalty of followers like Dolabella in "Shakespeare's Politics of Loyalty: Sovereignty and Subjectivity in *Antony and Cleopatra*," *SEL* 33 (1993): 343–63.

55. Jonathan Gil Harris, "Narcissus," 417.

56. Aristotle, *History of Animals*, trans. D. M. Balme, Loeb Classical Library (Cambridge: Harvard University Press, 1991), 539a, 18–26.

57. Aristotle, *Generation of Animals*, trans. A. L, Peck, Loeb Classical Library (Cambridge: Harvard University Press, 1943).

58. Quoted in Henry Harris, *Things Come to Life*, 5.

59. Ibid., 5.

60. Latour, *Pandora's Hope*, 154.

EPILOGUE: What about Bacon?

1. Deborah Harkness, *The Jewel House: Elizabethan London and the Scientific Revolution* (New Haven: Yale University Press, 2007), 214, argues against "the erroneous impression that [Bacon] was a scientific visionary articulating a wholly new approach to nature." Instead, she suggests that vibrant communities of experimentalists in London at the end of the sixteenth century pioneered a hands-on, empirical approach and that Bacon ignores this tradition in his writing. Michel Malherbe, "Bacon's Method of Science," in Markku Peltonen, ed., *The Cambridge Companion to Bacon* (Cambridge: Cambridge University Press, 1996), 75–98, argues that Baconian induction "has nothing

to do with, and even contradicts today's form of science" (75). Graham Rees, "Bacon's Speculative Philosophy," also in *The Cambridge Companion*, 121–45, notes that Bacon's theories about the structure of the universe were "rather strange" (121). Paolo Rossi, *Francis Bacon: From Magic to Science*, trans. Sacha Rabinovitch (Chicago: University of Chicago Press, 1968), argues that "Bacon should certainly not be seen as the inventor of modern science on the grounds that he discovered the inductive method" and that "modern science owes less to Bacon's empirical experimentalism than to Galileo's mathematical theories" (138). See Julie Robin Solomon, introduction to Solomon and Catherine Gimelli Martin, eds., *Francis Bacon and the Refiguring of Modern Thought* (Aldershot: Ashgate, 2005), 1–16, for an attempt to reconcile the differing views of Bacon held by scholars from different fields.

2. Rossi, *Francis Bacon*, esp. 135–51; see also Lisa Jardine, *Francis Bacon: Discovery and the Art of Discourse* (Cambridge: Cambridge University Press, 1974), esp. 17–75.

3. Barbara Shapiro, *A Culture of Fact: England, 1550–1720* (Ithaca: Cornell University Press, 2000), 107–12. She argues that "Bacon's combination of the historico-legal 'fact' of human action with the natural fact established by observation and experiment made it possible for his successors to apply a familiar legal technique of verifying events in the human world to natural phenomena" (109).

4. Jardine, *Francis Bacon*, 10, argues that Bacon did not have a deep knowledge of scholastic philosophy and that "it is in fact possible to account both for Bacon's 'Aristotelian' assumptions and for his 'anti-Aristotelian' polemic in terms of the content of the dialectic manual, and contemporary polemical discussions of dialectical method."

5. Francis Bacon, "*Novum Organum*, or True Directions Concerning the Interpretation of Nature," in *The Instauratio Magna Part II: Novum Organum and Associated Texts*, ed. and trans. Graham Rees and Maria Wakely, *The Oxford Francis Bacon*, vol. 11 (Oxford: Clarendon Press, 2004), 99 (aphorism 63).

6. *Historia Ventorum*, in *The Works of Francis Bacon*, ed. James Spedding, Robert Ellis, and Douglas Heath, *Philosophical Works*, vol. 2 (London: Longmans, 1876), 21, axiom 12, my translation. The Latin text reads: "Quid confaciant Meteora diversarum generum ad ventes? Quid Terrae Motus, quid Imbres, quid Concursus ventorum et invicem? Concatenata enim sunt ista, et alterum alterum trahit."

7. Francis Bacon, *The Advancement of Learning*, ed. G. W. Kitchin (London: J. M. Dent, 1973), 97.

8. Ibid., 29.

9. Ibid., 100.

10. Ibid., 91.

11. Ibid., 33.

12. Ibid., 104.

13. Rees, "Bacon's Speculative Philosophy," 122.

14. Ibid., 122.

15. Ibid., 122–25.

16. Ibid., 124–27.

17. Ibid., 129.

18. Ibid., 132. He gives as an example Bacon's reluctance to embrace Epicurean atomism because of its incompatibility with Christianity.

19. Bacon, *Advancement*, 1.4.5, p. 26.

20. Jardine, *Francis Bacon*, 69.

21. On dialectic and the commonplace book method, see Mary Thomas Crane, *Framing Authority: Sayings, Self, and Society in Sixteenth-Century England* (Princeton: Princeton University Press, 1993), 12–38.

22. Bacon, *Advancement*, 165.

23. On Bacon's use of "rhetorical division or partition" and its relation to Ramist "dichotomous division," see Jardine, *Francis Bacon*, 174–78.

24. Bacon, *Advancement*, 132.

25. Ibid., 126–27.

26. Jardine, *Francis Bacon*, 142–43.

27. Francis Bacon, *Novum Organum*, 85, English translation by Graham Rees and Maria Wakely, in *The Instauratio Magna Part II: Novum Organum and Associated Texts*, ed. and trans. Graham Rees and Maria Wakely, vol. 11 of *The Oxford Francis Bacon* (Oxford: Clarendon, 2004). This edition is in progress—it will eventually have fifteen volumes, but only six have been published (not in order). The general editor is Brian Vickers.

28. Ibid., 87.

29. Ibid., 213.

30. Antonio Pérez-Ramos, "Bacon's Forms and Maker's Knowledge," in Peltonen, *The Cambridge Companion*, 99.

31. Ibid., 105.

32. Ibid., 110–13. See also Reid Barbour, "Bacon, Atomism, and Imposture: The True and the Useful in History, Myth, and Theory," in Solomon and Gimelli, *Francis Bacon*, 17–43, which details Bacon's changing attitudes toward atomism throughout his career. Barbour argues that Bacon found atomism to be useful, even if he could not be sure the theory was true.

33. Bacon, *Advancement*, 2.7.1, p. 90.

34. Ibid., 90–91.

35. Ibid., 104.

36. Ibid., 101.

37. Craig R. Thompson, trans., *Literary and Educational Writings 2: De Copia/De Ratione Studii*, in vol. 24 of *Collected Works of Erasmus* (Toronto: University of Toronto Press, 1978), 639. See Crane, *Framing Authority*, 57–59, for an account of the bee metaphor in early modern humanist educational practice.

38. Bacon, *Advancement*, 125.

39. Bacon, *Novum Organum*, 152–53. In Latin, Bacon contrasts "empirici" with "rationales."

40. Barbour, "Bacon," 17.

41. Jardine, *Francis Bacon*, 121.

42. Pérez-Ramos, "Bacon's Forms," 103.

Index